BUILD AND PROGRAM YOUR OWN
LEGO® MINDSTORMS® EV3 Robots

Marziah Karch

800 East 96th Street,
Indianapolis, Indiana 46240 USA

Build and Program Your Own LEGO® MINDSTORMS® EV3 Robots

ISBN-13: 978-0-7897-5185-0

ISBN-10: 0-7897-5185-2

Library of Congress Control Number: 2014955375

Printed in the United States of America

First Printing November 2014

Trademarks

All terms mentioned in this book that are known to be trademarks or service marks have been appropriately capitalized. Que Publishing cannot attest to the accuracy of this information. Use of a term in this book should not be regarded as affecting the validity of any trademark or service mark.

LEGO® and MINDSTORMS® are registered trademarks of The LEGO Group.

This book is not authorized or endorsed by The LEGO Group.

Warning and Disclaimer

Every effort has been made to make this book as complete and as accurate as possible, but no warranty or fitness is implied. The information provided is on an "as is" basis. The author and the publisher shall have neither liability nor responsibility to any person or entity with respect to any loss or damages arising from the information contained in this.

Special Sales

For information about buying this title in bulk quantities, or for special sales opportunities (which may include electronic versions; custom cover designs; and content particular to your business, training goals, marketing focus, or branding interests), please contact our corporate sales department at corpsales@pearsoned.com or (800) 382-3419.

For government sales inquiries, please contact governmentsales@pearsoned.com.

For questions about sales outside the U.S., please contact international@pearsoned.com.

Executive Editor
Rick Kughen

Development Editor
Ginny Bess-Munroe

Managing Editor
Sandra Schroeder

Project Editor
Seth Kerney

Copy Editor
Paula Lowell

Indexer
Cheryl Lenser

Proofreader
Jess DeGabriele

Technical Editor
John Baichtal

Publishing Coordinator
Kristen Watterson

Interior Designer
Mark Shirar

Cover Designer
Mark Shirar

Compositor
Mary Sudul

Contents at a Glance

Table of Contents

About the Author

Marziah Karch enjoys the challenge of explaining new gadgets and complex technology to beginning audiences. She is the author of several books, including *Android Tablets Made Simple*. Her writing has appeared in *Wired* magazine, About.com, and the GeekMom blog on Wired.com.

Marziah is a senior instructional designer for NWEA in Portland, Oregon. She holds a master's degree in Instructional Design and is working on a Ph.D. in Library and Information Management. When she's not feeding her geek side with new gadgets or writing about technology, Marziah enjoys life in the Pacific Northwest with her husband and two children, all of whom are LEGO enthusiasts.

Dedication

This book is dedicated to Pari and Kiyan. Keep on building.

Acknowledgments

I'd like to thank Melissa Kelly for her photos, robot club attendance, and enthusiasm. Ada and Jay also get credit for helping. I hope they build amazing robots. Harold spent countless hours helping me build every single one of those demo robots. Travis Coon over at LEGO Education/Pitsco was amazingly helpful with demos and previews and suggested resources. Finally, I'd like to thank the wonderful editorial staff at Pearson for everything they did to bring this book to press.

We Want to Hear from You!

As the reader of this book, *you* are our most important critic and commentator. We value your opinion and want to know what we're doing right, what we could do better, what areas you'd like to see us publish in, and any other words of wisdom you're willing to pass our way.

We welcome your comments. You can email or write to let us know what you did or didn't like about this book—as well as what we can do to make our books better.

Please note that we cannot help you with technical problems related to the topic of this book.

When you write, please be sure to include this book's title and author as well as your name and email address. We will carefully review your comments and share them with the author and editors who worked on the book.

Email: feedback@quepublishing.com

Mail: Que Publishing
         ATTN: Reader Feedback
         800 East 96th Street
         Indianapolis, IN 46240 USA

Reader Services

Visit our website and register this book at quepublishing.com/register for convenient access to any updates, downloads, or errata that might be available for this book.

Introduction

If you've been looking for a fun introduction to robotics without having to solder wires or learn advanced programming languages, the LEGO MINDSTORMS EV3 is just the ticket. You can make and program robots using a graphical interface and LEGO interlocking parts. When you're ready for a new challenge, you can hack the operating system and use more advanced languages such as Java. You can also connect EV3 robots to harness the combined computing power or have EV3 robots communicate wirelessly with each other.

When you're ready to get more social with your projects, there are First LEGO Robotics Leagues, LEGO robotics clubs, and LEGO robotics–themed camps. You don't even have to be a kid to enjoy playing with LEGO robotics. I once helped build a team robot at a Google-sponsored booth at the SXSW Interactive festival in Austin, Texas. Part of the challenge even included hacking the Android phone app used as a remote controller for the robot. There wasn't a child in attendance, yet everyone was as excited as a kid in a candy store.

The EV3 is such a wonderful kit for every age, not only because it's a solidly built toy that contains everything you need to get started, but also because you don't have to stick with just the items in the box. The Cubestormer 3 is a world-record-setting Rubik's Cube solver built mainly out of EV3 parts and a Samsung Galaxy S4 phone. One creative 12 year old used the power of an EV3 to build a relatively inexpensive braille printer. Check out Chapter 12, "Extending Play," for more details.

As you can see, the EV3 goes beyond what one could traditionally expect out of a toy. On top of creative play, it offers some great opportunities for problem solving, engineering, and learning while having fun.

This book is intended to help get you started. The projects are all suitable for new users of all ages, whether in a classroom or going solo. Wherever possible, this book explains the why as well as the how. Read the book, tear apart the projects, and improve upon them. There's absolutely no reason why your floor-cleaning robot can't also send you an email to let you know when the floor is clean—or climb stairs.

As you go through the book, because failure can teach you some things that success cannot, you'll occasionally find projects that do not work on the first try. This is mostly intentional, but don't worry—I do explain what went wrong and how to make it right. It's all part of the learning process that, when you've gone cover to cover, will help you become a better builder.

Two versions of the EV3 are available for purchase. Those buying from a toy store will probably have the EV3 Home Edition, whereas those ordering for First LEGO Robotics League or a classroom will tend to have the LEGO Education edition. Don't worry—this book has you covered on both fronts.

What's in This Book

Chapter 1, "What's in the Box?": This chapter goes through the parts and pieces in the EV3 Home Edition, including the included sensors, motors, and test track.

Chapter 2, "What's in the LEGO Education Box?": This chapter goes through the parts and pieces in the LEGO Education set. Even if you don't have this set, you can separately purchase a lot of the parts, so it's a good overview and might give you expansion ideas.

Chapter 3, "Comparing the EV3 and NXT": If you've played with the previous version of LEGO MINDSTORMS, you'll want to check out the differences and improvements in the EV3.

Chapter 4, "Building Your First Bots": This chapter goes over the demo robots available from LEGO and offers a little more insight into the things you should watch out for as you make them.

Chapter 5, "Building the LEGO Education Bots": This chapter goes over the demo models for the LEGO Education set. Build everything from a self-balancing robot boy to a spinning top factory.

Chapter 6, "Hacking What You Have": There's no need to reinvent the wheel when you get started. Take what you learned from the demo models and use it to make something new.

Chapter 7, "Make Your First EV3 Program": This chapter takes the robot you built in Chapter 6 and shows you how to make your first program. This chapter also demonstrates that there are many ways to make the same program.

Chapter 8, "More MINDSTORMS Programming: The Line-Following Robot": This chapter goes more in depth into programming. You'll learn about variables and flowcharting and hopefully gain a little insight into thinking like a programmer.

Chapter 9, "Engineering the Floor-Cleaning Robot": In this chapter, you'll learn how to make an autonomous robot that self-navigates and avoids collisions while cleaning your floor.

Chapter 10, "The Color Magic Card Trick": Rather than making a vehicle, this chapter focuses on the difficult engineering task of getting the robot to deal and identify cards by color.

Chapter 11, "Daisy-Chaining Projects": In this chapter, you'll hook two EV3 robots together and see how they can communicate. You'll also explore wireless communication between EV3 robots.

Chapter 12, "Extending Play": This chapter explores how to install leJOS, an alternative operating system for Java programmers. You'll also look at robotics clubs, robot decoration, and compatible parts from other vendors.

Appendix, "Glossary": The appendix is a glossary of some of the more unusual words you might find in this book.

How to Use This Book

Throughout the book, you'll run across notes and tips.

TIP

Tips are useful pieces of information that will help you avoid a problem or be more efficient.

NOTE

Notes are extra bits of information about the subject. They might mark some great places to study later.

What's In the Box?

Did you know LEGO makes a programmable robot? Of course you did. That's why you bought this book. The EV3 is the third generation of MINDSTORMS, replacing the NXT 2.0. This version has a smarter processor, new sensors, and lots to love. In Chapter 4, "Building Your First Bots," I'll go over a few of the robots you can build right away with the EV3 home edition, but for now we'll start looking at the items that come in the box.

Figure 1.1 shows the EV3 retail box. A separate LEGO Education edition is also available, which I discuss in a little more detail in Chapter 2, "What's in the LEGO Education Box?" Both the LEGO Education and retail versions of the EV3 use the same programmable Brick and building techniques. They just differ in parts.

FIGURE 1.1 The LEGO MINDSTORMS EV3 retail package contains everything you need to build a variety of robots.

NOTE

Before you open your box, think about where to store the pieces. In this chapter, I suggest using a clear plastic storage bin and recloseable plastic bags. Also, the retail EV3 kit requires six AA batteries and three AAA batteries.

Unboxing **MINDSTORMS EV3**

Open the LEGO MINDSTORMS EV3 box from the sides to remove the box contents, but be careful not to destroy the box in your eagerness. The box itself is important for the EV3, because the outer box wrapper is actually a testing track in disguise. Carefully cut the outer box wrapper along the dotted line on the back (see Figure 1.2).

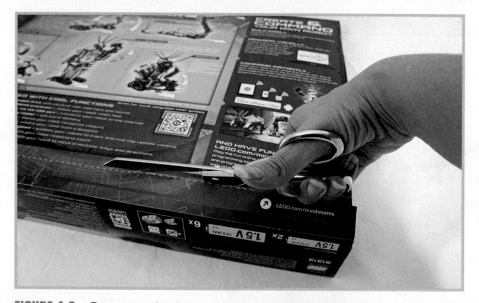

FIGURE 1.2 Cut open the box wrapper as indicated.

Cut any tape securing the sides of the wrapper to the main EV3 cardboard box, and then unfold the wrapper to reveal your snazzy new testing track (see Figure 1.3).

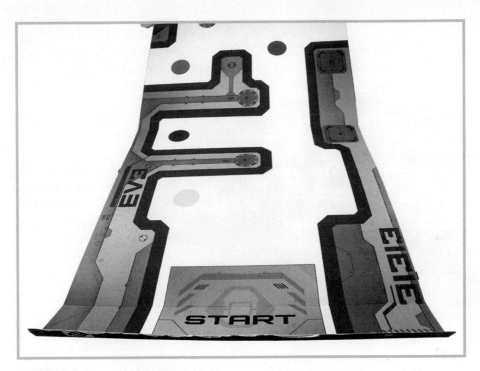

FIGURE 1.3 Unfold the test track.

Now that you have both an open box and a testing track, you can start unpacking your box. Your new EV3 is awesome, but the box doesn't leave enough room for storage. When you dump out the box's contents you'll find several bags, a pamphlet, and some stickers, as shown in Figure 1.4. You can't reuse the bags as storage after you open them, and the EV3 home edition box isn't big enough to hold an assembled robot.

FIGURE 1.4   Here are the bagged pieces and parts from the EV3 box.

Before you start ripping apart those bags to get to your EV3 pieces, I suggest getting a clear plastic storage bin with a lid and several sizes of recloseable plastic bags to store the parts (see Figure 1.5).

Hardware stores also have a variety of small parts organizers that work well as MINDSTORMS parts organizers. You might even want to buy two storage options—one for part organization and one for any assembled or partially assembled robots. Assembled robots can get quite large, so consider something at least as large as a milk crate. You may want to flip through the projects in Chapter 4 to get an idea of the size range for assembled robots.

FIGURE 1.5   Here's a simple organization system using plastic bags and a large plastic box.

SORTING THE PARTS

There's no perfect way to sort your LEGO pieces into bags or storage boxes. Sorting by size, type, or color are good starting points. I find that a combination approach works best for me. I put all straight beams in the same bag or compartment, no matter the color, but sort my pegs by color. Beams with bends are hard to sort and pull out when I need them, so I separate them by size and shape.

If you don't know what I mean by "beam" and "peg," don't worry. I take you on a tour through all of these parts later in this chapter.

As you use your kit, you'll find the sorting style that works best for you. Just make sure you have plenty of bags and a sturdy box on hand to start sorting them. The surest way to lose your pieces is to start playing without any sorting system in place.

Speaking of those parts, one of the first things you might notice as you pull them out of the box is that they're not shaped like traditional LEGO bricks. This is because the EV3 pieces are part of the LEGO Technic family, which uses a system of interlocking pins and gears for stronger and more elaborate builds.

The LEGO Family Tree

If you grew up playing with LEGO bricks, the MINDSTORMS set might look very different to you, and the two kinds of pieces are mostly incompatible. To understand why standard LEGO pieces don't easily snap together with the EV3 pieces, you need to meet some of the LEGO family.

LEGO DUPLO

The large LEGO DUPLO blocks are intended for toddlers and children under age 6 who might swallow smaller pieces or have a hard time fitting them together. DUPLO blocks are easy to grip and assemble. You can use a LEGO System building plate underneath a DUPLO building, but due to their large size, DUPLO blocks are mostly incompatible with other LEGO lines.

LEGO System Bricks

Older children move on from DUPLO blocks to traditional LEGO System bricks. LEGO markets these bricks to certain age ranges, so kits start out as simple toys with easy instructions and later evolve into complicated sets with multiple books of building instructions. LEGO System bricks are what most people picture when someone says the word *LEGO*.

LEGO System bricks come in a wide variety of themes for building everything from superheroes to city dwellings, and even a LEGO Friends line marketed specifically to girls is available. However, all of these pieces for all of these different themes are compatible with each other. You can use pieces of Boba Fett's spaceship to build a home for your Hobbits and decorate it with flowers from LEGO Friends. This was a major plot point in *The LEGO Movie*.

The bricks for both LEGO System and LEGO DUPLO fit together using studs (the bumps on a LEGO brick). The incompatibility between LEGO System and LEGO DUPLO is just in the size of the studs. LEGO Technic, which I cover next, is different. Most of the construction takes place without studs.

LEGO Technic

LEGO Technic items are designed for kids around age 10 and up. The LEGO Technic line is an assembly system composed of pegs, rods, gears, and pulleys that emphasize movement and mechanics. Some LEGO Technic pieces even come with motors and remote controls. For the most part, assembly happens without any studs or traditional LEGO System bricks.

This is where LEGO MINDSTORMS comes in. Although technically the MINDSTORMS line is separate from the LEGO Technic line, the pieces are generally compatible with each other and use the same stud-free method of construction with pegs and gears. If ever you need an extra pneumatic lift or tractor wheel for your EV3, you can shop for Technic parts and find that most of them are compatible. Sometimes you may even find a few Technic parts inside of LEGO System sets.

You can find a few studs here and there in LEGO Technic, but those are mostly for things such as the "lights" on a truck or other decorative elements. The lack of studs allows for easier assembly and stronger and more flexible structures. It does, however, require a little more planning for your structures.

Core LEGO Units

Technic family LEGO pieces have a unique form of measurement. You might have noticed that the parts are labeled with measurements such as "3," and LEGO instructions in general avoid using a lot of text descriptions. This practice makes for easier international releases, but sometimes you can get frustrated when all you have is a picture of a product with 3. Three what?

You might assume that because LEGO is a Danish product, the measurement refers to metric units. That isn't the case. Figure 1.6 shows basic axles against metric measuring tape. The axle on the right is a size 3.

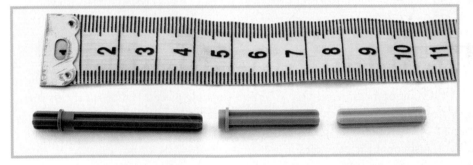

FIGURE 1.6   You can see that LEGO units are not perfectly metric.

The measurements don't quite align with any other metric units, so a 3 length of axle can't be 3 meters, centimeters, or millimeters long. In fact, it's a little more than 2 centimeters long. A 3 length of axle is actually three *beams* long, as shown in Figure 1.7.

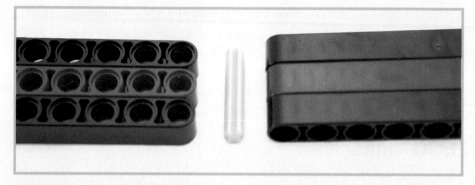

FIGURE 1.7   The best measure of a Technic part is the beam unit.

LEGO units are measured against themselves, no ruler required. The unit is one LEGO Technic beam unit. Beams measure the same width and depth, and the holes along beams are distributed at the same interval. A size 6 beam has six holes in it. All you ever need to measure the size of your axle is a beam. However, the paper instructions included in your EV3 kit also include a 1:1 size comparison of many axles for reference as you build.

Let's get started exploring the most common types of LEGO pieces compatible with your EV3 kit.

Beams

Beams and pegs are your most basic LEGO Technic building pieces. As just discussed, the beam is the basic unit of measurement for your EV3 building pieces. Figure 1.8 shows the basic long beams. The retail EV3 set comes with four size 15 beams, four size 13 beams, four red size 11 beams, and eight size 9 beams. If you're ever in doubt about which size beam you have, just count the peg holes.

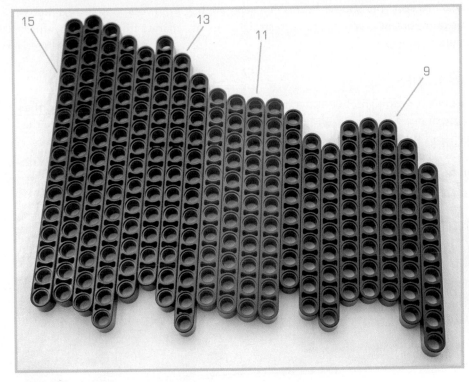

FIGURE 1.8 Various straight beams in red and black.

Straight beams range from size 15 down to size 3. The EV3 home edition contains twelve size 3 beams, ten size 5 beams, and six size 7 beams. Figure 1.9 shows a size 3 beam.

FIGURE 1.9 The straight beam.

You might want to sort your shorter beams in a separate bag from your long beams, but I find the most important practice is to keep straight beams separate from angled beams, which I discuss next.

Angled Beams

Beams come not only straight, but also angled. Figure 1.10 shows what I like to think of as the sled shape of beam, but it is really known as the double-bent lift arm. You'll notice the little cross shapes for axles on each end. I cover those later in this chapter when I discuss axles. For now, the important thing to know is that you can connect beams with angles to your design to avoid having corners be your weak spot.

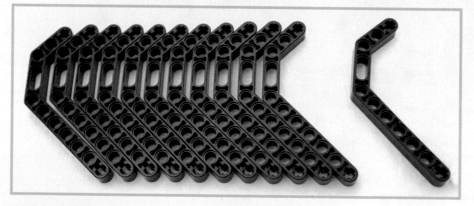

FIGURE 1.10   Notice the two 45-degree bends and a longer arm on one side.

The EV3 home edition contains 12 of this particular double-bent lift arm, so feel free to make heavy use of them in your designs.

The kit also contains 12 of the single-bend lift arms with a 45-degree angle bend, as shown in Figure 1.11. I like to store my double-bend beams separately from my single bends, just because all those bends tend to hook other pieces, which makes pulling them out of the bag harder.

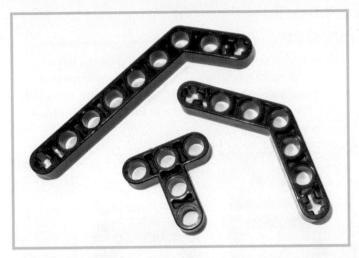

FIGURE 1.11 These beams have single 45-degree bends.

Angled beams are not limited to 45-degree angles or to beams with only offset bends (see Figure 1.12). The kit also contains four of the smaller 45-degree angle lift arms, eight and six of two different 90-degree angle beams, and even four beams with a T-shape. You'll have plenty of strong, bendy shapes for building arms, legs, and other structures that need strength.

FIGURE 1.12 Two more types of angled beams.

Beam Frames

Figure 1.13 shows one other class of beam, the rectangular beam frame, of which there are two types. In one type, the ends of the rectangle's long sides extend past the edges of its short sides, and the other type forms a typical rectangle. These beams are important because they not only enable you to make stable structures with bends, but also to vary the angle of the connecting holes, so you can use these beams (you get two of each kind) to connect other beams in six different directions.

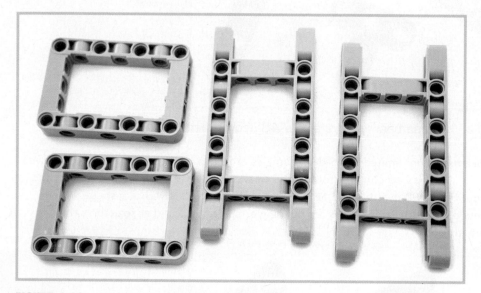

FIGURE 1.13   Use beam frames to connect other beams in different directions.

Pegs

How are you going to connect all these beams? With pegs, of course. Pegs are versatile connectors, because their round shape means they move. Connect two beams using one peg, and the beams can rotate with the peg as an axis. Use two pegs in different locations, and the beams stay stable and rigid.

The basic black peg is the easy go-to peg for these tasks. Figure 1.14 shows the black pegs. The EV3 kit contains 95 of them. Black pegs are one beam deep on either side, so if you use a black peg to secure two beams together, they will be flush. They also have a slight texture to them, so they provide some friction against free motion. The joint is still moveable, just not as much as that offered by a gray peg.

FIGURE 1.14 Black pegs can secure two beams.

I suggest keeping the black pegs in a separate bag, so you always have a simple peg available. They come packaged together, so no sorting is involved.

In addition to the bag of black pegs, the kit also has a bag of mixed color pegs and bushings, as shown in figure 1.15. Let's talk about the pegs, first. There are gray, single pegs (similar to the black pegs), of which the kit has four. The gray pegs do not offer the light friction of black pegs, so use gray pegs in intentional joints and moving parts.

FIGURE 1.15 Various pegs, axles, and bushings.

You'll also find red, longer pegs with connectors on one side (ten) and four beige pegs with two beam lengths on one side. Those pegs are good for either adding distance or attaching two beams at once. If you don't like beige, 38 of the same shape are available in blue.

Some of the items shown in Figure 1.15 aren't strictly pegs. The kit also contains 28 blue half pegs/half axles and 12 red axles (described in the following section).

Finally, Figure 1.16 shows double pegs, some of which have axle bushings or extra beam connections. The parts have names like "module bush" and "cross block." These crossbeam parts are great for stabilizing beam connections.

FIGURE 1.16   Module bushes and cross-blocks.

Axles

Figure 1.17 shows a variety of axles you can find in your EV3 home edition. Axles connect pieces together, but unlike pegs and beams, single axles can connect two pieces in a way that does not allow them to move. The important features of an individual axle include the length and the location of stops, if any.

FIGURE 1.17   Axles come in multiple sizes.

If an axle has no stop, the connected piece could potentially slide off. You'll need to either use other pieces to reinforce the connection or add a bushing at the end.

TIP

Sometimes small axles get stuck in other pieces. You can use another axle to poke it out; however, a properly sized Philips screwdriver also does the trick.

Bushings

Bushings are connections that go on the end of axles and can be used to stop an axle. You get 11 yellow half bushings and 9 red full bushings (see Figure 1.18). The half refers to the width of half a Technic beam unit. You can use half bushings to prevent an axle from stopping, and use full bushings for either that purpose or to connect two axles in a pinch. For a better connection, you should use axle connectors.

FIGURE 1.18 Use bushings and half bushings on the ends of axles.

Axle Connectors

You use axle connectors (see Figure 1.19), as you would expect, to connect two axles together. You can use them to make two axles work as one long straight axle, or use them to give axles a 45-degree angle or add beam connections for pegs. These parts are also known as angle elements.

FIGURE 1.19   Use red angle elements to connect axles. (Callouts indicate the number contained in each set.)

In addition to pegs, beams, axles, and bushings, the kit contains additional parts that combine those pieces. Figure 1.20 shows a half beam, half bushing that allows for peg connections perpendicular to axle connections. Figure 1.21 shows more variations of bushings, beams, and pegs.

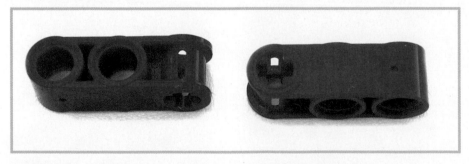

FIGURE 1.20   These parts are also called cross blocks.

FIGURE 1.21 Use these gray parts as steering elements and levers.

Ball Joints

Ball joints, shown in Figure 1.22, are meant to fit into sockets and rotate freely, just like the ball joints you have in the bones in your shoulders and hips, allowing a better range of motion in your arms and legs. Ball joints also resemble the tow balls that trucks use to haul trailers. Because they can be used anywhere you need a part to move within a socket, they're also known as tow balls. Your EV3 ball joints, or tow balls, come with peg or axle connectors.

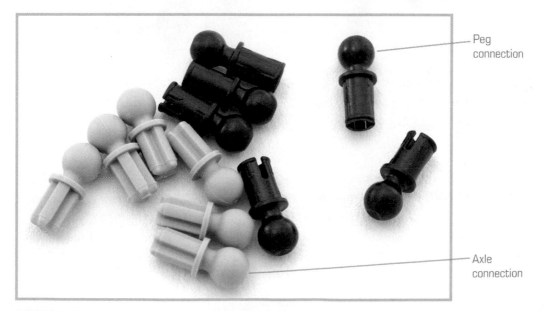

Peg connection

Axle connection

FIGURE 1.22 Tow balls come with friction pegs or axles.

Figure 1.23 shows steering links that can be connected using ball joints. Steering links have round connections on either end, so you could put ball joints on both sides to attach a trailer or other object.

FIGURE 1.23   Here are two sizes of steering element.

Gears

Now that you've seen some of the basic connecters in the LEGO Technic series that are in your kit, let's look at the parts that give some leverage and motion to your projects.

Figure 1.24 shows the various gear types found in your LEGO EV3 set, which includes four-point gears, as well as wheel gears with teeth in various sizes. Notice the gear centers. You can put a peg or axle off center in many of them to make a crank.

FIGURE 1.24   Here are the various gears that come with your EV3.

Figure 1.25 shows the worm gear. You can drive this gear with another gear. It is especially useful for making cranks, lowering or raising cord, or making the arms of bridges.

FIGURE 1.25   The oddly named worm gear is useful for making cranks.

Figure 1.26 shows the two cams included in the kit that you can use to make items with piston actions.

FIGURE 1.26   Cams have various axle slots.

Your EV3 comes with a large variety of wheels and wheel treads, as shown in Figure 1.27. You can use the wheels with or without the treads. In fact, you can swap your tire treads out for tank treads, depending on the sort of robot you want to make.

FIGURE 1.27 The EV3 offers many different wheel options.

The tank treads shown in Figure 1.28 are unique to the retail EV3 set. The LEGO Education EV3 set has interlocking hard plastic pieces instead of the rubber tank treads. These tank treads are strong and flexible, but there's no way to make them larger or smaller for alternate uses. Your kit comes with two.

FIGURE 1.28   The EV3 home edition comes with rubber treads.

In addition to the tire treads and the tank treads, there's also a single red flexible band. It comes in a white paper box, as shown in Figure 1.29. If you're a LEGO fan, this is like the packaging LEGO brick system pieces use for minifigure capes. Feel free to discard the box, but keep the band. It can be useful as either a tire tread or as tension between pieces or gears.

FIGURE 1.29 Here's a band after it has been removed from the box.

Figure 1.30 shows the ball parts: three red balls, a ball holder, and a ball gripper. These particular ball parts are unique to the EV3 home edition, and they're a slight change from the rainbow balls of the NXT robotics kits.

Balls are mainly used for shooting and target practice. They're typically stored inside the holder and then fired individually from the gripper. For a demonstration, see the models in Chapter 4.

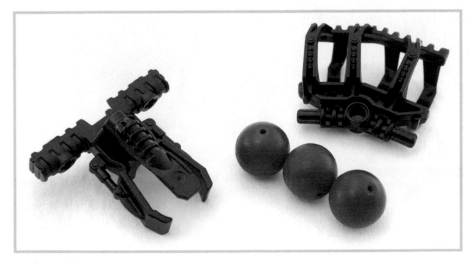

FIGURE 1.30 The balls, the ball gripper, and the ball holder.

Parts with Flair

Some EV3 parts are not really structural as much as they are for decoration or flair—not that these parts are never important to the structure of a robot. The kit contains wings, swords, and dials to round out your robots.

Wings

Figure 1.31 shows the small and large wings that come with the retail EV3 set, and they offer a place where you can put some of your EV3 stickers. The set comes with three pairs of small 3x7 and three pairs of large 3x11 wings (six wings each size). The wings have beam connections on the bottoms and sides, and they can support some weight, although in most cases they're just used as fins, wings, or puppy-dog ears. LEGO officially calls these parts beams with bows.

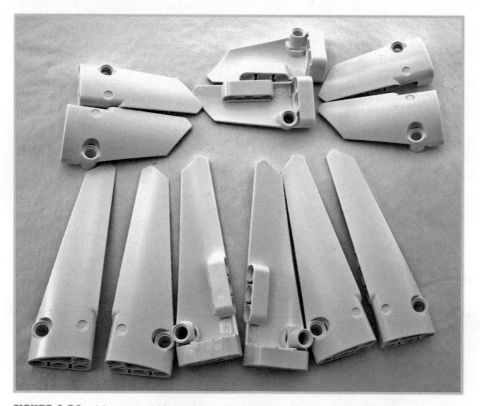

FIGURE 1.31   Here are the wing pieces or beams with bows.

Figure 1.32 shows two cousins of the wings that you can use for things such as the edges of a tank or combined for a face-like shape for your robot. These two corner pieces also get EV3 stickers.

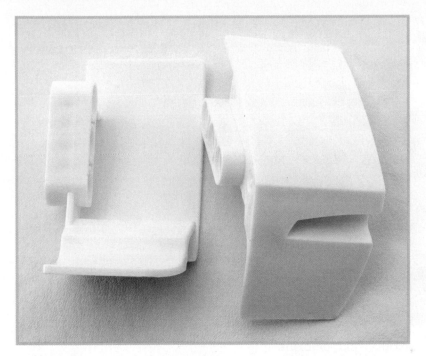

FIGURE 1.32   The modeling elements are also called car parts.

Spikes

Figure 1.33 shows spikes or dials. These parts do not come in the core LEGO Education set, but for retail kit owners, they're fantastic for jazzing up a robot. You get four white dials and six red ones. These are also sometimes called bions.

FIGURE 1.33   The spikes, dials, or bions.

Other Decorative Parts

Figure 1.34 shows the six pieces in the home edition that I think of as swords. They could also work as insect legs or spikes, but these pieces are mostly decorative and come with an axle connection on end.

FIGURE 1.34 Here are the swords.

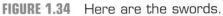

Figure 1.35 shows the four batwing-like shapes that round out the decorative flair pieces. LEGO calls these blades. These are half a beam long and generally rotate freely. The batwings are used as spinning tank parts in the very first model included in the EV3 instructions.

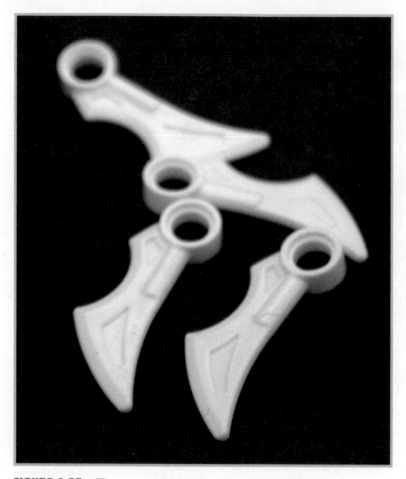

FIGURE 1.35   These could be batwings or blades.

The Brains and Brawn

The discussion so far has centered on some of the parts that you can use as the foundation for your robot, but not the parts that make it think or the motors that make it move. Let's cover those important parts now.

EV3 Intelligent Brick

Figure 1.36 shows the EV3 Intelligent Brick, which represents the brains of your robot. It's the heaviest part in the set because it's packed full of batteries and computing parts, which allow it to do all the thinking and powering for everything else. You load all programs onto the Brick and you can even program directly from the Brick itself.

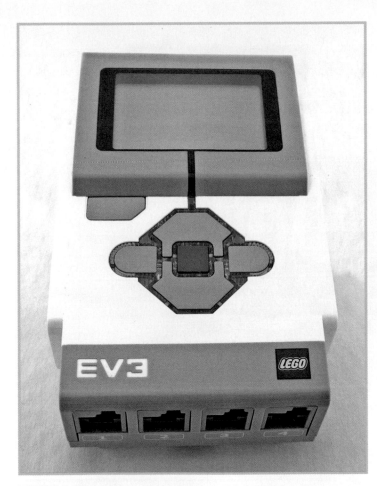

FIGURE 1.36 The brains behind every EV3 robot.

Notice the EV3 Intelligent Brick has plugs labeled 1–4 on one side and A–D on the other. These spots are where your cables go to power sensors and motors. The Intelligent Brick also has beam connections on the sides and bottom. Before you get too far along, let's make sure your Intelligent Brick has power.

You can remove the back of the Intelligent Brick and insert six AA batteries, as shown in Figure 1.37.

FIGURE 1.37   The batteries loaded in the back.

You'll probably go through a lot of batteries with your EV3, depending on how often you use it. Investing in rechargeable batteries might be a good idea. However, a lot of MINDSTORMS enthusiasts notice that when you use rechargeable batteries, your robot might need to have those batteries recharged more and more often. You might also find your robot acting a bit sluggish over time. If you're entering your robot in a competition, going with the less environmentally friendly disposable batteries on game day is probably worth your while.

Now look at the sides of your Intelligent Brick a little closer.

Figure 1.38 shows the speaker side, where your programmed robotic sounds can be broadcast. There are identical 3×3 L-shaped beam connections on both long sides of your Intelligent Brick.

FIGURE 1.38   Here's the speaker side of the Intelligent Brick.

Figure 1.39 shows the opposite side from the speaker, which has the USB and SD slots. You can use the SD slot to insert an SD flash memory card that contains programs or data. You can use the USB connection to connect your EV3 to your computer to transfer programming from the desktop app. We'll start discussing the desktop app in more detail in Chapter 7, "Make Your First EV3 Program."

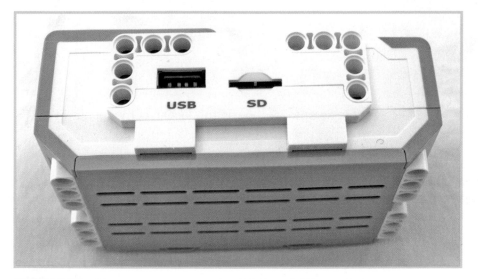

FIGURE 1.39   Here you can see the USB and SD card slots.

Now that you've examined the Intelligent Brick, take a look at the motors and sensors that it powers.

Servos

Figure 1.40 shows the two large servos that ship with the EV3 retail kit. These servos can power wheels, arms, or other large items. The red parts rotate in a circular motion, and you can connect pegs around the edges of the face or an axle in the center. In addition, the servos have several other spots for making axle or peg connections to other pieces.

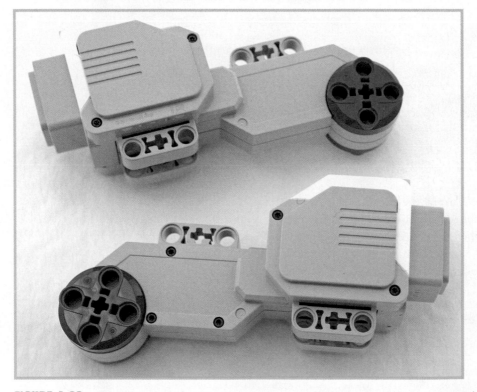

FIGURE 1.40   The large motor servos can power the larger components in your kit.

Figure 1.41 shows the medium motor servo. This servo only spins the middle axle connection, although it has peg connections available near the moving part and you can use them to connect gears that are powered by the servo.

FIGURE 1.41   The medium motor servo has a connection for the middle axle.

Sensors

Figure 1.42 shows the touch sensor, of which only one comes in the EV3 home edition (the NXT had two). This sensor detects when the red end is depressed. It can detect collisions when you put it on the end of a car, but it is also extremely useful as a switch to switch a robot on or off or make it reverse directions or reset. The Gyro boy robot in the LEGO Education kit uses the touch sensor as a reset switch.

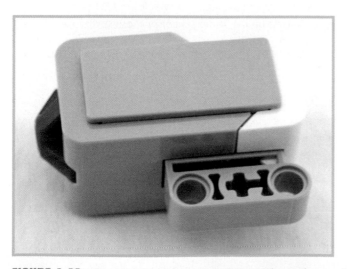

FIGURE 1.42   The touch sensor senses when the red area is pressed.

The color sensor, shown in Figure 1.43, detects light. It can sense different colors and differentiate between shades of light and dark. One sensor comes in the kit. You can use it to sort items, detect their color, or make a robot that follows a line. Chapter 7 and Chapter 8, "More MINDSTORMS Programming: The Line-Following Robot," go into more detail on uses for the color sensor.

FIGURE 1.43   The color sensor can detect both ambient and reflected colors.

Figure 1.44 shows the infrared sensor and remote (or beacon), which are exclusive to the home edition. The LEGO Education set substitutes a sonic sensor without a remote. The infrared sensor and remote (which uses AAA batteries) are easy to adapt for making just about any car or tank robot into a remote-controlled device. You can also use the infrared sensor without the remote to detect whether anything is in front of the sensor. One sensor and remote come in the kit.

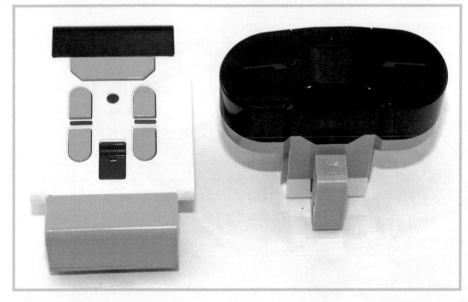

FIGURE 1.44   Here is the infrared sensor with the beacon.

Cables

Now that you have looked at your Brick, servos, and sensors, take a look at the cables needed to connect them.

Figure 1.45 shows the two types of cable that are included with the EV3 kit. One is the USB cable, which is like a standard USB printer cable found just about anywhere. You use it to connect the Brick to your computer to transfer programs. The other cable is the proprietary cable used to connect servos and sensors to the Brick. It's similar to CAT 5 cable that you would use to connect your computer to an Ethernet router, but its wiring is different and you cannot replace it with a generic Ethernet cable.

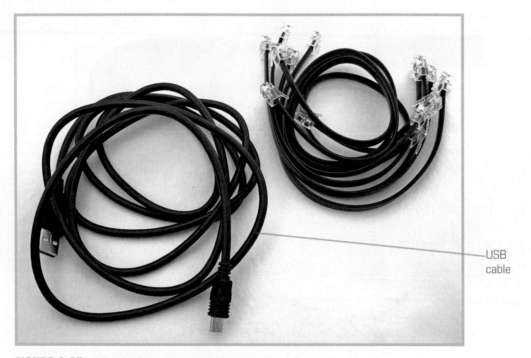

USB
cable

FIGURE 1.45   Your kit comes with a variety of cables.

You get an assortment of both long and short cables. There are four 25cm cables, two 35cm cables, and one 50cm cable. When you build a robot, try to select the shortest cable that will do the job. Otherwise, you'll spend a lot of time winding the cables to get them out of the way. You may also want to label the ends of the cable with electrical tape and magic markers to avoid confusion.

Finally, the retail kit comes with a set of stickers (see Figure 1.46) that can go on your wings and corner parts. An instruction book, shown in Figure 1.47, shows you how to build your first robot. It's actually three robots in one.

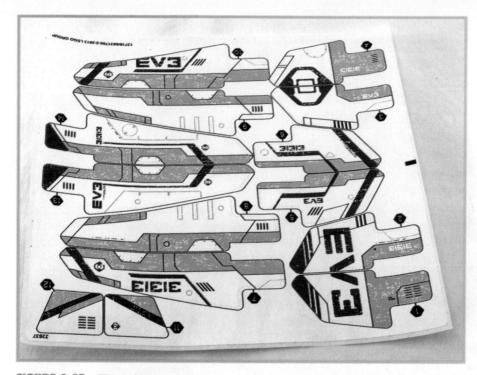

FIGURE 1.46 The sheet of stickers is an optional decoration for wing elements.

After you have unboxed all your parts (including the box and the test track that comes with it), you can open that instruction book and start building your very first robot or press on to Chapter 2 for a look at the EV3's education edition.

FIGURE 1.47   If you lose the manual, you can still download instructions for this robot.

Summary

In this chapter, you opened the MINDSTORMS EV3 kit and explored the various parts that ship with it—beams, pegs, axles, bushings, gears, ball joints, sensors, the EV3 Brick—and even the box itself. You also learned about the LEGO family tree, considered brick storage options, and prepped your EV3 for play by adding batteries.

2

What's In the LEGO Education Box?

Chapter 1, "What's in the Box?" covered the EV3 home edition, and I touched on the fact that it isn't the only version of the EV3. This chapter covers what's in the LEGO Education edition. Combined, Chapters 1 and 2 can give you clarity about which version to purchase if you're still pondering, and show you what the other version looks like if you've already made up your mind.

This chapter does not go over every single part included with the LEGO Education kit, because many of them are the same as those included in the EV3 home edition. Instead, this chapter focuses on the differences between the two kits and also covers the EV3 expansion set available through LEGO Education. Let's get started.

ABOUT LEGO EDUCATION

LEGO Education North America is a partnership between LEGO and Pitsco Education, an educational product company founded in 1971. The companies formed a joint venture in 1997. LEGO Education offers slightly different versions of LEGO products that are geared toward educational settings. Rather than buying LEGO Education products in stores, you can order them through the LEGO Education USA website http://www.legoeducation.us/.

Storage Box

The first thing you'll notice about the LEGO Education version is that it comes with its own plastic storage box, as shown in Figure 2.1. When you open the clear plastic lid, you'll find a cardboard insert that has the name of the kit on one side and a complete parts inventory on the other, as shown in Figure 2.2. No test track comes in this kit, unlike the home edition, where part of the cardboard box becomes the test track. There's also no cardboard box to throw away.

FIGURE 2.1 The Education edition comes in a handy box.

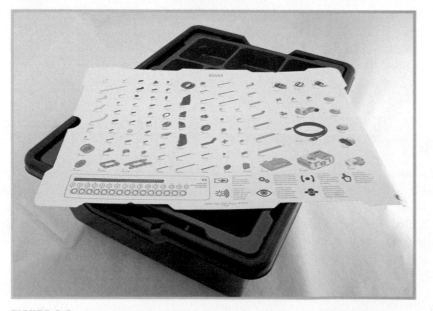

FIGURE 2.2 The cardboard insert includes a parts list.

Underneath the cardboard insert is a red tray with divided sections for sorting small parts or storing accessories, as shown in Figure 2.3.

FIGURE 2.3   There's a built-in sorting tray.

Lift that out to see the basic parts for your EV3 kit, bundled into non-reusable plastic bags and a few cardboard boxes. I recommend sorting these parts into plastic sandwich bags instead of relying on the red sorting tray. Not only does it not have enough room to hold all the pieces, it's also the part most likely to tip and spill pieces all over your floor when you try to carry the kit to the table without the lid secured.

NOTE

You should invest in a large rubber band, a length of ribbon, or a Velcro tie system to keep the box lid closed during transport. The Education kit box is great for sorting and organizing your pieces, but the lids do not snap closed securely enough to keep your contents from spilling if you drop the box or even hold it upside down.

Color Scheme

One of the differences you might notice right away between the home edition and LEGO Education versions of the EV3 is that they have different color schemes. The home edition has a strict red, black, and white color scheme with stickers for the white wing parts to make them look like well-worn robot or spaceship parts. The LEGO Education set has green, blue, and yellow parts mixed in with the red, black, and white. There are no stickers, and the wings are black.

Figure 2.4 shows the multicolored size 3 beams in the LEGO Education kit. Although most beams are black, white, or red, these smaller size 3 beams can add some pops of color to your designs.

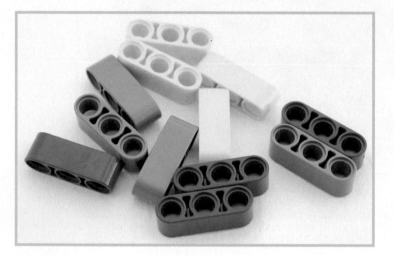

FIGURE 2.4   There are four each of these brightly colored size 3 beams.

NOTE

Remember that LEGO Technic pieces are measured in standard beam-sized units. Each hole in a standard beam is one unit apart from the next hole, and the beam itself is one unit deep. If you are ever stuck trying to figure out the size of a piece, just pull out a standard beam and measure.

Figure 2.5 shows the wing parts (also known as *panels*) that ship with the core Education set. The wing parts on the LEGO Education set are black rather than white, and they do not have decorative stickers. Functionally, they are the same as in the home edition, although the core Education set ships with fewer of the pieces than the LEGO home edition. There are only four total making one pair each of small and large. Adding stickers may be problematic for educational settings where the stickers could potentially be peeled off with multiple student users.

FIGURE 2.5   There are very few black wing panels in the core set.

Ball Caster

The LEGO Education set includes two pieces that might appear a bit odd if you're not familiar with them (see Figure 2.6): a metal ball bearing and a female receptor for it. These two pieces make up the ball caster. It's not designed as a ball-shooting device like the parts that come with the home edition set. Instead, it is useful as a stabilizer for robots that need to move along a flat surface.

FIGURE 2.6   The ball and caster makes for a great stabilizer on flat surfaces.

Figure 2.7 shows how the ball caster is meant to fit together. Try holding on to the beam end of the ball caster and rolling it along a smooth surface. Notice how easily it glides.

FIGURE 2.7   This assembled ball caster offers a smooth glide.

TIP

After you install it, the ball caster is unlikely to fall out by accident, which is good news for your builds. If you ever want to remove the ball bearing from the caster set, just poke it from the back with an axle.

Tank Tracks

The LEGO Education set doesn't have any rubber tank treads/tracks. Instead, LEGO has included 54 interlocking pieces that you can snap together to make your own tank tracks, up to 54 tread lengths long (see Figure 2.8).

FIGURE 2.8   You assemble your own treads in the LEGO Education edition.

Although you give up the simplicity of the rubber track pieces with this kit, it offers the flexibility for you to have one giant track for an assembly line or three separate tracks for a tricycle tank. Tank track pieces are actually a feature in many other Technic building sets.

Rechargeable Battery Pack

Your LEGO Education EV3 is slightly greener than its retail counterpart. Rather than use standard AA batteries, you can opt to use the rechargeable battery pack that comes with the kit, as shown in Figure 2.9. You're not locked to one method or the other, so use whichever one makes sense for your build. Some users complain that the rechargeable battery has a little less oomph when used for some tasks, so keep that in mind as you build.

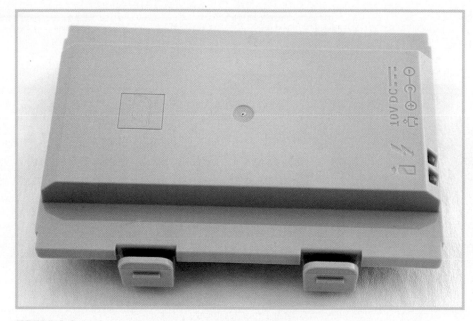

FIGURE 2.9   The rechargeable battery back.

Some earlier versions of the battery were defective, so if you open your box and find that your battery won't charge, contact LEGO Education to see whether they can replace it.

Sensors

Like the home edition, the LEGO Education set contains two large motor servos and one small motor servo. The rest of the sensors are different from what you would find in the home edition.

> **NOTE**
>
> The Intelligent Brick is identical between the Education and home edition. Programs that run on one set will run on the other, and the default demo program ships with both. You can always buy additional sensor parts or pieces a la carte and add them onto your MINDSTORMS kit.

Touch Sensors

You get an extra touch sensor with the Education set, as shown in Figure 2.10. Instead of just one, now you have two. Put them on either end of a robot for collision detection or use them as buttons for different functions.

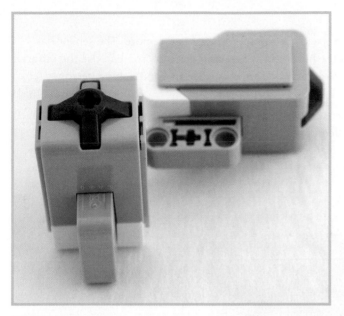

FIGURE 2.10   Two touch sensors are included with this kit.

Gyro Sensor

The gyro sensor, shown in Figure 2.11, detects motion, direction, and angle to keep a robot balanced and sense which direction it is pointing. This usage is particularly impressive in the Gyro Boy build, which I discuss in Chapter 5, "Building the LEGO Education Bots."

FIGURE 2.11   The gyro sensor helps robots maintain balance.

Sonic Sensor

The LEGO EV3 home edition comes with an infrared sensor and remote. The sensor has slit-like "eyes" and could serve as the head of any humanoid robot. You can use the infrared sensor to avoid collisions or to make the robot interact with an approaching person. However, this sensor doesn't come in the Education set.

Instead, it has a sensitive sonic sensor, which detects sound. You can use it for collision avoidance as well, which is also demonstrated by the Gyro Boy model. It also has an eye-like appearance, as shown in Figure 2.12, so your humanoid robot won't have to go without a face.

FIGURE 2.12 The sonic sensor—notice how the "eyes" are different from the home edition's infrared sensor.

This sonic sensor should also be familiar to MINDSTORMS NXT 2.0 owners, because it is a beefed-up version of the sensor that shipped in both the Educational and home edition versions of that robotics kit.

Gears

In addition to the other minor differences discussed so far, the LEGO Education set has a few more interesting gears, such as the gear with beam holes (see Figure 2.13), which is actually part of a turntable gear. I cover it further during the expansion set discussion "Gears and Joints" later this chapter.

FIGURE 2.13   The upper part of the turntable gear.

The beam with boss and pin or steering knuckle arm in Figure 2.14 resembles a winding handle. On one side is a peg that could fit into a beam or be flipped to form a handle for something cranking an axle.

FIGURE 2.14   The steering knuckle arms resemble a winding handle.

To round things out, the Education kit also includes some flexible double bushings made out of a rubbery material (see Figure 2.15).

FIGURE 2.15   These double bushings are also called dampers.

Overall, the LEGO Education core set provides some interesting parts for robot building as requested by the classroom customers.

LEGO SOFTWARE

The LEGO Education set does not ship with software. That isn't unusual because the home edition doesn't ship with software, either. You just download it off the website.

You can download the home edition software and use it freely on the LEGO Education set. There's a caveat, however. The home edition software includes instructions for models that you can't construct with the pieces contained in the LEGO Education set.

LEGO Education does *sell* its own educational version of the software for $99 per seat. It's designed to run in classrooms (or homeschools or clubs or groups) where a teacher wants to monitor the activities of the students. It includes lesson plans for the instructor; expansion packs can be purchased separately for more activities. The LEGO Education version of the software includes instructions for models that can be made with just the core set or the addition of the expansion set.

The Expansion Set

What comes with the LEGO Education version of the EV3 is known as the *core set*. It's a good set, but a great upgrade for it is available that you should get if you can afford it. The LEGO Education Expansion Set for the EV3, shown in Figure 2.16, runs an extra $99 but comes with a lot of fantastic parts you can use right away. The LEGO Education version of the software (which is another $99 and sold separately) also includes instructions for building additional figures with this expansion set.

FIGURE 2.16 Opening up the expansion set.

The expansion set comes in another stackable box, the same size and style as the EV3 kit, only with a white sorting tray instead of red. The set also includes a parts list when you flip the cardboard insert under the cover. As you can see, the box is full of extra parts. You don't get an extra programmable brick, but you get a lot of parts to extend the building possibilities. I also recommend using sandwich bags to sort the extra parts after you open the bags, as well as an extra rubber band, ribbon, or Velcro fastener to keep this lid on, too.

Extra Wheels

Some of the first things you might notice in the box are the wheels (see Figure 2.17). Some wheels are so large that they weren't even inside a sealed plastic bag with the other small parts.

FIGURE 2.17 This kit has an amazing selection of wheels.

The EV3 Education core kit comes with only enough wheels to make a wheeled robot, but the expansion set adds some variety. Not only are there ten more treads and 22 more tire rims, but they also range in various sizes from large to small. The expansion set even allows you to make a robotic elephant that uses the extra wheel rims (sans treads) sideways as feet.

Extra Beam Frame Elements

The square beams that come with the EV3 Education set are fantastic elements for engineering strong structures. The problem is you only get two of each. The expansion set adds a significant number to the pile (see Figure 2.18), which allows you to make much larger, more stable robots.

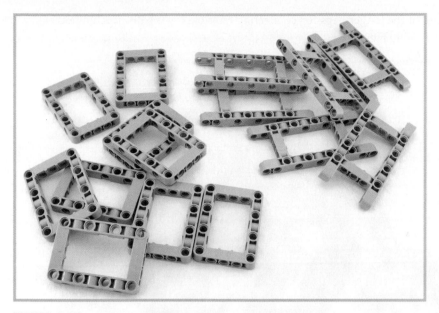

FIGURE 2.18   The expansion set has enough beam frames to build larger structures.

Yes, that also means you get extra pegs (see Figure 2.19).

FIGURE 2.19   There are 170 extra peg connectors in the expansion set.

You also get more of the longer pegs for connecting multiple beams or attaching other accessories, as shown in Figure 2.20.

FIGURE 2.20 The expansion set includes 14 beige and 22 blue connectors.

You also get more cross blocks that combine pegs with axle and beam connections, as shown in Figure 2.21.

FIGURE 2.21 Cross blocks and beams with snaps.

Some of the cross blocks found in this kit are specialized and not found in either the LEGO EV3 home edition or in the core Education set. The black pieces shown in Figure 2.22 are three-pin steering hubs.

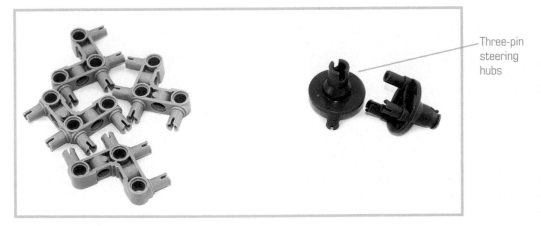

Three-pin steering hubs

FIGURE 2.22 Angular beams with snaps and three pin steering hubs.

The expansion set comes with a few extra standard straight beams, but it also comes with interesting beams, such as the triangle thin type shown in Figure 2.23.

Triangle
thin type
beams

FIGURE 2.23   Extra angled beams.

The expansion set also has thin straight beams and a large pile of toggle joints, as shown in Figure 2.24.

FIGURE 2.24 Catch with cross-holes.

Gears and Joints

The expansion set includes a lot of items beyond basic building and connecting with pegs and axles. Robots need to move, and the expansion set provides more possibilities that facilitate movement than the core set. For example, it offers additional gears, as shown in Figure 2.25.

FIGURE 2.25 Extra gears add extra possibilities.

These are basically the same sorts of gears you would find in other sets. LEGO takes it up a notch and also provides two gear racks, as shown in Figure 2.26, so you can have gears that move something up and down (or side to side). You might make a robot that gets taller to reach objects or create a scanner that moves a sensor side to side.

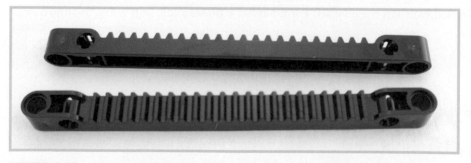

FIGURE 2.26 Gear racks provide for lifting and moving.

In this case, the gear rack also includes an axle and peg connection on the ends, providing multiple ways to secure it to your robot.

There's also a differential gear, which can help cars turn corners. When you use a differential, you generally place it between two wheels to allow one wheel to turn faster than the other (the sort of thing that happens when turning corners). The expansion set only comes with one differential gear, as shown in Figure 2.27.

FIGURE 2.27   The differential gear helps your builds turn corners.

You get one differential, but you also get two turntables (see Figure 2.28). Yes, it looks like four pieces, but two pieces actually fit together and form a turntable gear that allows both sides to spin independently. This is also something that can be useful for creating robots with smooth steering motions.

FIGURE 2.28 These parts assemble into turntable gears.

Speaking of smooth motions, the expansion set includes two spare ball cups for your ball bearing (see Figure 2.29). No spare ball bearing comes with the set—sorry. One cup is exactly the same as the one that comes in the core LEGO Education set. The other is more like a ball joint and adds a larger beam connection.

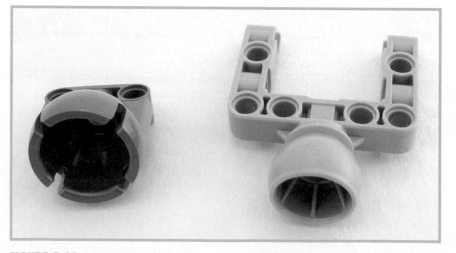

FIGURE 2.29 Extra ball cups for your ball caster

Speaking of ball joints, the expansion set includes extra ball joints and steering beams, as shown in Figure 2.30.

FIGURE 2.30 Extra tow or ball joints and steering beams.

The set also has two "universal joints" that connect two axles and allow a short range of motion along one axis, much like a knee joint on your leg (see Figure 2.31). They're mainly used in motors to transfer motion to a gear at an angle.

FIGURE 2.31 Universal joints bend and connect with axles.

The two odd-looking velocity joints included in the expansion set, and shown in Figure 2.32, look like little maces for mini-figures.

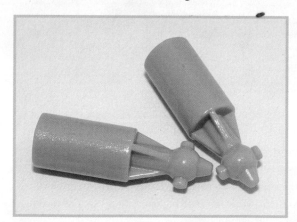

FIGURE 2.32 These are velocity joints—not medieval weapons.

They're not maces, though. Velocity joints couple with the velocity receptors, shown in Figure 2.33, to make what is known as a constant velocity or CV joint. This joint is another option for advanced motors.

FIGURE 2.33 Catches and velocity receptors (cardian cups).

Tank Cleats

You might notice a bunch of flexible red pieces in your expansion set, like those shown in Figure 2.34. These 28 pieces are actually extra traction for your tank treads. If you ever wondered why each piece of tank tread had two holes in it, wonder no more.

FIGURE 2.34 28 rubber tank cleats for tank treads.

The LEGO expansion set doesn't come with many extra tank tread pieces at all. In fact, there are only five extra pieces. Their inclusion might only be to let you see what those red tank cleats look like when you put them on the treads, as shown in Figure 2.35.

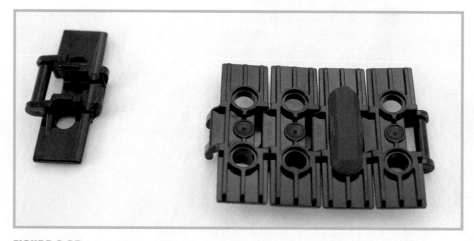

FIGURE 2.35   The five extra pieces of tank tread with one of the 28 cleats.

Wings and Other Decorative Pieces

The LEGO Education core EV3 set doesn't come with many wing pieces compared to the home edition. The expansion set makes up for that with more wing pieces and more sizes of wing pieces than are found in the home edition (see Figure 2.36). The color scheme sticks with the mostly black wings, but you do get a few small red wings.

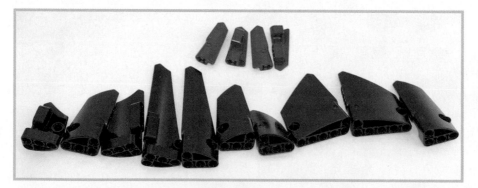

FIGURE 2.36   More wing (panel) options.

There are also more car parts in the expansion set, with four large bumper-like pieces that you can use to build vehicles and other items (see Figure 2.37).

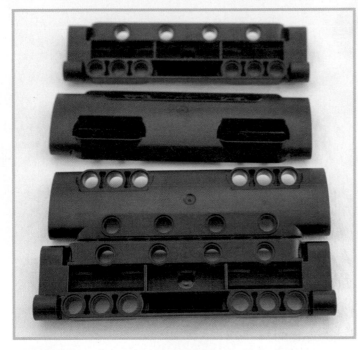

FIGURE 2.37 The "bowed panels" or car parts.

In addition to the car parts and wings, the expansion set also has more dials and includes a few LEGO system pieces, such as the round buttons shown in Figure 2.38.

FIGURE 2.38   LEGO System style buttons.

The expansion set also includes the blue tubes shown in Figure 2.39. Their purpose is decorative, but they're not included in the EV3 home edition set, so your robot will have a more unique look with this touch.

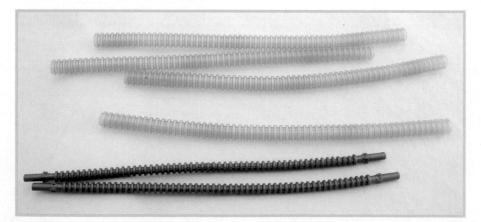

FIGURE 2.39   The blue tubes.

To go along with the blue decorative tubes, the kit also provides blue, turquoise, and yellow decorative parts so you can create some truly unique robots (see Figure 2.40). These are also parts that do not come in either the Education core or the EV3 home edition. The red part in the picture is a driving ring and does serve a more functional purpose.

FIGURE 2.40 More options for decoration and expansion

The last bit of extra parts in the expansion set includes the both decorative and functional rubber band set (see Figure 2.41). You get eight total rubber bands in four different sizes (four sets of two). The rubber bands follow the red, white, yellow, and blue color scheme of the other decorative parts and come in little cardboard boxes like you might have also seen with capes in LEGO system pieces.

FIGURE 2.41   Extra tension rings in their boxes.

Summary

This chapter offered a look at the LEGO Education version of the EV3 and the parts that are unique to that set. You also learned about the expansion set for the EV3, which is sold through LEGO Education. The expansion set provides a lot of unique pieces and items that allow for more complicated robots, but it also includes a lot of decorative pieces exclusive to that set. LEGO Education sells a separate version of the EV3 software, but the home edition version is also still an option.

3

Comparing the EV3 and NXT

You might have an NXT 2.0, or see one for sale on eBay. You might wonder whether you should upgrade or what you could do with your old NXT 2.0 now that you live in an EV3 world. Can you combine the two systems? This chapter answers those questions.

THE ORIGINAL MINDSTORMS

In 1998, LEGO released the first version of the LEGO MINDSTORMS, the RCX. It used a yellow programmable Brick that was compatible with both some Technic pieces and LEGO System pieces. The brick itself was somewhat less compatible with sensors, however, which used a different connection system.

Although current software doesn't officially support RCX, you can use an older version of ROBOTC to program it, which you can download from www.robotc.net/download/rcx/.

You might find old RCX units on eBay. I found several RCX Bricks listed for around $30, and entire sets for $75. Early sets were plug-in models, but later sets were battery powered.

The NXT Versus the NXT 2.0 Versus the EV3

LEGO released the original NXT in 2006 (see Figure 3.1), and followed it up with a slight refresh in 2009: the NXT 2.0. The biggest difference between the two systems was mainly the parts that were included in each kit. LEGO upgraded the software between the NXT and NXT 2.0, and if you ran a program designed for the NXT 2.0 on your NXT Brick, it automatically upgraded the firmware to remain compatible with it. There was no difference in the physical appearance of the Intelligent Brick.

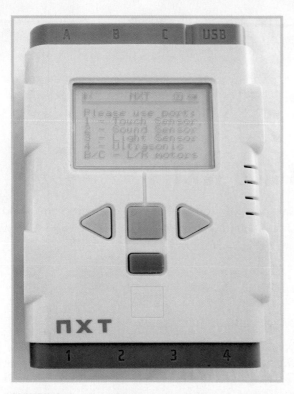

FIGURE 3.1 The NXT 2.0 Intelligent Brick was a notable upgrade over the original version.

For all intents and purposes, there are only three MINDSTORMS Intelligent Bricks: RCX, NXT, and EV3. Because the RCX hasn't been on the market in a while and is less compatible with current hardware and software, this chapter focuses on the NXT and EV3.

TIP

You can still find several websites, instructions, and tons of great books on NXT and NXT 2.0, such as *Basic Robot Building with LEGO NXT 2.0*, also available from QUE Publishing

---

As of this writing, there were still a few stores selling new NXT 2.0 models, although they started labeling them as "collectable," and the eBay used listing price was still in the $250 range. The EV3 is more expensive than the NXT 2.0, new or used, so what is so much better in this set that justifies the extra expense? Well, you don't *have* to choose. Not right away. The NXT system is compatible with EV3, and it should be fully supported until at least 2015, but let's go over the advantages.

The Brick

The main difference between the EV3 and NXT systems is the upgraded Intelligent Brick. On the EV3, the Brick has a faster processor, better on-the-Brick programming, and a Linux-based operating system. Figure 3.2 shows the front of both Bricks.

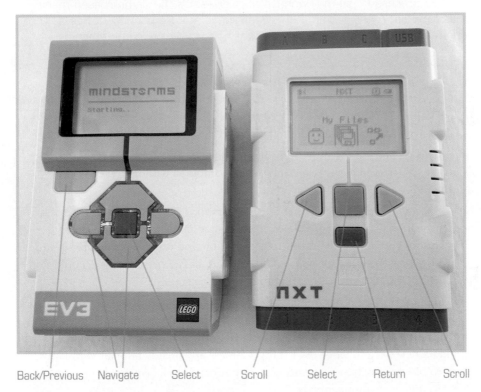

Back/Previous   Navigate   Select   Scroll   Select   Return   Scroll

FIGURE 3.2   Compare the Intelligent Bricks side-by-side.

Figure 3.2 shows a side-by-side comparison of the EV3 and NXT Bricks. You can see that although both are about the same size, the EV3 has a slightly larger screen. The EV3 has up and down buttons along with the Back button (on the upper-left corner for the EV3, and center bottom for the NXT). The NXT has only two directional buttons, a center (Select) button, and a Back button below it. You absolutely need those extra buttons on the EV3, because you have more programming options (discussed later in this chapter in the "Programming" section).

When you power on the Bricks, the EV3 buttons are backlit, making it possible to find them in the dark. That said, the screen is not backlit so the net benefit here is minimal.

Behind the scenes, the EV3 has a lot more processing power than the NXT. The EV3 does have one penalty with the newer system, however, and that is in boot speed.

Why is the boost in processing important? That powerful ARM-9 processor and Linux-based operating system offer a lot of advanced programming potential, some of which I talk more about in Chapter 7, "Make Your First EV3 Program," where you learn about LEGO's partnership with LabVIEW that resulted in an enhanced desktop programming experience.

Although you could go beyond the desktop programming system with the NXT 2.0, using alternatives on the EV3 is even easier, especially when combined with the rest of the hardware on the EV3.

With the NXT 2.0, you could string together a few commands on the Brick itself (to test a robot and make sure it was running, for example). You could not do any advanced programming without connecting the robot to a computer and downloading a program. If you didn't want to use the visual programming software LEGO provided, you could find alternatives like ROBOTC, a proprietary programming language.

The EV3 allows full on-the-bot programming, not just a few commands for testing. It also makes using ROBOTC and other alternative programming environments easier. Although LEGO never hid the code—you can download the NXT firmware code as open source—it still required learning a new system. Since many programmers are already familiar with Linux, LEGO's choice to use a Linux-based operating system for the Intelligent Brick encourages even more programming environments.

Sensor and Motor Connections

The LEGO MINDSTORMS NXT and EV3 use the same proprietary phone cord-like connector cable (also know as the RJ12), which makes them mostly compatible with each other. On the bottom of both Bricks, as shown in Figure 3.3, you can see that both Bricks have four sensor slots, labeled 1–4. You can only connect sensors to these ports, even though you use the same cable to connect sensors and motors.

FIGURE 3.3   NXT and EV3 both have four sensor ports.

Despite the fact that the four ports you see here are visually the same, the new EV3 sensors do not work on NXT Bricks. You can go the other direction and use your old NXT sensors, but the old NXT firmware doesn't support the improved EV3 sensors.

NOTE

The improvements to the EV3 sensors include the following:

- The color sensor can sense seven colors instead of six.
- The sonic sensor (LEGO Education kit) can now be used as sonar.
- The gyro sensor (in the LEGO Education kit) is more sensitive to rotation.
- The infrared sensor is also more sensitive and accurate.

Next, let's compare the motor ports, as shown in Figure 3.4.

FIGURE 3.4   Notice that the EV3 has one more motor port than the NXT.

The differences are immediately visible from the top of the brick. The NXT has only three motor ports (A, B, and C) whereas the EV3 has four (A, B, C, and D). That means you can put 25% more things that move on your EV3. You can also see a minor difference in USB ports. The NXT uses the Type B USB plug that's most commonly associated with older printers and scanners. The EV3 uses a mini USB connector like those commonly found on digital cameras.

You can also see, from the angles shown here, that the EV3 is just slightly taller than the NXT because the screen sticks further out than the rest of the Brick. The two Bricks are otherwise very close in size, which is good news for anyone who wants to adapt an NXT build to work for EV3.

NOTE

The mini USB connection on the EV3 is different from the style of USB connector currently used for most mobile phones chargers, which are usually micro-USB connectors. Fortunately, mini-USB is still plenty common and it's easy to get a replacement part if you need one. Just make sure not to confuse it with the aforementioned micro-USB.

The Sides

Figure 3.5 shows the speaker side of both Bricks. The only major difference for the LEGO robotics fan is the placement of the peg holes on the side of the Bricks. The EV3 is positioned so that beams can be connected more toward the end of the Brick. It's a small change, but it makes for more stable robotic structures.

Speaker

FIGURE 3.5   Both EV3 and NXT Bricks have speakers on the side.

Figure 3.6 compares the other side of the Bricks, and you can see a real difference. In addition to the peg placement, the EV3 has another USB connection and an SD card slot. The NXT has no extra connections on this side at all.

Having an SD card slot allows you to expand your robot's memory and load files and assets instead of storing everything on the Intelligent Brick itself.

FIGURE 3.6 The SD slot and USB port are new additions to the EV3 Brick.

If the NXT already has a USB slot on the bottom, why does it need another on the side? Allowing you to connect a USB dongle to the side of your bot might permit you to run not-yet-invented add-ons, and it can help you chain multiple Intelligent Bricks together.

TIP

The USB port on the side also allows you to connect bots together if you have more than one EV3. That's right—you can daisy chain up to four EV3s together. The Bricks will communicate with each other for even more advanced bot design. That's something you definitely cannot do with an NXT.

As an example, you may want to make a robot that plays piano. You could add more robotic fingers by chaining Intelligent Bricks together and using them to power more servos.

Programming

For the NXT, your programming options are limited to things you can do on the desktop and transfer over to the Intelligent Brick. You can do some basic commands on the NXT and order it to run a motor or test a sensor, but this isn't really programming. On the EV3, you can actually program on the bot, a feature that opens up a host of possibilities. Figure 3.7 shows the on-screen programming capabilities of both models.

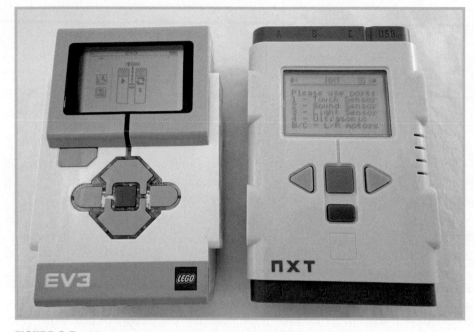

FIGURE 3.7   Here you can see the programming screens.

As you can see, the extra EV3 buttons are necessary in regard to programming and debugging existing programs. That isn't to say that EV3 on-the-bot programming is easy or natural. It's still cumbersome, but at least it is possible.

> **NOTE**
>
> The new EV3 desktop software is compatible with the NXT, although some sensors and features do not work because they're not supported on the NXT hardware. Feel free to download the free EV3 home edition software and use it to program your NXT. It means less of a learning curve if and when you do decide to upgrade your old bot. You can also save your old programs for use with an EV3 set.

Advanced and Alternative Programming Environments

You can use the desktop programming environment with both the NXT and EV3 systems, but you can also use other programming languages and environments. Both the NXT and EV3 are open source to encourage a growing hacker community.

NOTE

Open source means that the software code is released to everyone, and anyone can freely use or modify it. The Linux operating system is open source, and modifications of it include everything from computers to watches to phones. You can download the EV3 source code from GitHub at https://github.com/mindboards/ev3sources and modify it in any way you choose. Traditionally, open source programmers who do something cool post their updated source code back to the community to allow everyone to enjoy it.

As I am writing this, there aren't a lot of EV3 programming environment options, but that doesn't mean they aren't on the horizon. Here are the ones available now for NXT:

- **Enchanting**—Enchanting is a visual programming environment based on MIT's Scratch. You can download it from http://enchanting.robotclub.ab.ca/tiki-index.php. As of this writing, Enchanting is available for NXT but not for EV3.
- **ROBOTC**—ROBOTC was developed at Carnegie Mellon for use with the LEGO competitor, VEX Robotics. It is available from http://www.robotc.net. As of this writing, ROBOTC is available for NXT (and RCX) but not EV3, although an EV3 version of ROBOTC is in the works.
  ROBOTC, as the name implies, is a C-based programming language, which makes it a great choice for high school and advanced students or programming classrooms. It also has the advantage of being a single language to use with multiple robotics platforms. ROBOTC isn't free, however. Licenses start at $49 per year.
- **Other Languages**—NXT-compatible adaptations of Ruby, Lua, Ada, C, and many other programming languages exist that aren't available for the EV3—yet. Until they are, if you're into advanced programming, you still have some good reasons to hold onto your NXT.

NOTE

As of this writing, the EV3 now supports a version of Java called leJOS. You can also download an alternate version of the EV3 OS called EV3Dev from https://github.com/mindboards/ev3dev. EV3Dev uses Debian Linux and allows easier support for Python, bash/dash, Awk, Perl, Lua, and Ruby. Don't worry. You don't have to uninstall your existing OS to use this. You can load EV3Dev through your SD card. You just need to restart your Intelligent Brick without the SD card installed to restore the Brick.

Part Compatibility

NXT and EV3 both use LEGO Technic parts and they're 100% compatible with each other in regard to beams, pegs, and axles. Mixing and matching parts between the NXT and EV3 sets isn't a problem.

> **TIP**
>
> The EV3 has a lot more bent beams, and the NXT has a lot more straight beams. Between the two, you can get quite the collection.

Sensors

EV3 sensors are more advanced than NXT's. Some, like the infrared sensor, simply weren't part of the NXT set. As mentioned earlier in this chapter, you can't use EV3 sensors on your NXT. You can, however, use NXT sensors on your EV3. You might also be able to use RCX sensors if you have a converter cable, although I don't know of anyone who has tried this yet.

Motors

The motors have a slight difference in appearance, but NXT and EV3 motors and servos are compatible with each other. The NXT came with three large servos, so if you combine an NXT and EV3 set, you could hook up a large servo to every motor port on the EV3.

Batteries

Figure 3.8 shows the backs of the EV3 and NXT Bricks. Both Bricks have integrated beams in the same position, which is handy for converting builds intended for one device to another. They also both take six AA batteries each. However, that's where the power simi-larities end. The NXT and EV3 have separate rechargeable battery pack adapters (the EV3 adapter ships with the LEGO Education set) and these adapters are not compatible with each other. If you find a battery adapter on eBay, make sure you verify which device it fits.

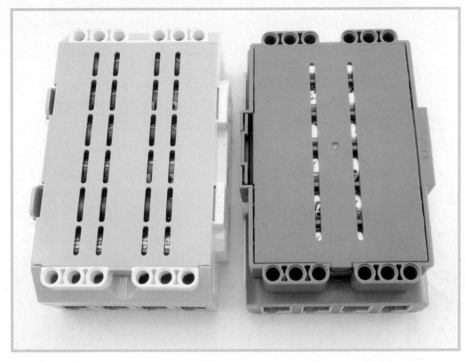

FIGURE 3.8   The back of the Intelligent Bricks have integrated beams.

Summary

The NXT might not be the current MINDSTORMS model, but it is still a decent robotics option. EV3 desktop software runs on the NXT, and many EV3 and NXT parts are compatible with each other, though not all of them. The NXT is still supported by a vibrant hacker community and probably will be for many years in the future. If you upgrade from your NXT to the EV3, you might want to hold on to the parts to enhance your EV3.

4

Building Your First Bots

One of the fantastic things about the EV3 is that great instructions are already available for building all sorts of robots. This chapter looks at the currently available instructions for the LEGO MINDSTORMS home edition set, and Chapter 5, "Building the LEGO Education Bots," explores the instructions available for the LEGO Education set.

You can use this chapter as a reference to see whether you want to try a project, or you could challenge yourself to build along. On top of building along, you can program along. You can choose to download and use the finished EV3 programs as is, or you can follow the mission instructions and build the programs yourself as you go.

Downloading Instructions

Your EV3 home edition set comes with the instructions for the Track3r bot in the written manual. If you lose the printed manual, the Track3r and all other basic instructions are available online at http://www.lego.com/en-us/mindstorms/products/starter-robots/.

You can also get the instructions by downloading the EV3 Home Edition software and launching the "missions," or by using the tablet app available for iPads and Android devices.

> **NOTE**
>
> If you choose to run the included program for your EV3 bot, you should run the mission from within the EV3 desktop software rather than downloading the finished program. On some bots, the downloaded software gives errors for missing blocks.

If you have a tablet, you can enjoy some fantastic 3D building instructions courtesy of LEGO and Autodesk, as shown in Figure 4.1. The LEGO MINDSTORMS 3D Builder app is available either through the Google Play Store or the Apple App Store. The app works on tablets running iOS or Android, but it does not work on phones.

Instructions are shown with 3D animations, and you can spin the model around to see the whole picture. The tablet app is especially helpful for new builders who might be unfamiliar with LEGO's style of instructions.

FIGURE 4.1   When building using the LEGO MINDSTORMS 3D Builder app, each step is animated and rotatable.

The EV3 Starter Robots

Let's go over the basic builds. All the starter robots substitute a 3 for an e somewhere in the name, and all of the bots' instructions are broken up into separate missions to either build the bots in stages or add program features.

These pauses between missions allow you to test your robot as you go, since sometimes building a robot from someone else's roadmap can make it confusing to troubleshoot mistakes. The smaller missions also enable you to see how small changes can make big differences in the purpose and function of different robots.

Track3r

The Track3r robot instructions ship with your EV3 Home Edition, and the program to drive your Track3r is already installed on your bot, whether it's the home edition or LEGO Education version. It is the demo program. You can take advantage of that in later chapters by building tanks you can test out before doing any of your own programming. You can also follow along and program your bot to complete the other missions. The programs are relatively simple at this stage and teach you how the programming interface works.

You can build the Track3r in five stages (missions), running the demo software after each build to see how it reacts differently.

TROUBLESHOOTING

Is your bot not moving? The two things you should check when something goes wrong at this stage are your battery levels and port connections.

Low batteries make a bot sluggish, and connecting a sensor or motor to the wrong port means that the program won't tell it to move.

Mission 1

You can see the end-build of Mission 1 in Figure 4.2. The Track3r has blades on one side that spin and an infrared head that doesn't do much other than provide it with good looks. If you launch the demo, you'll see the bot look from side to side by showing different eye graphics, make noise, and drive by itself.

It doesn't go very far. This demo program really is just designed to show you that you made a robot that can run. Hold onto that thought, though. I like to use this same demo program to test other robots I build. It doesn't have to just be a tank bot. I'll show you how this works in Chapter 7, "Make Your First EV3 Program."

FIGURE 4.2 The completed Mission 1 robot.

If you unfold the cardboard test track from your EV3 box (see Figure 4.3), you can actually use this blade and driving motion for a demonstration on your track. Place a tire on the marked area of the track, and the blades will knock the tire around. The robot is self-propelled in this case, and it doesn't vary in its pattern. This is the "mission" part of the first mission. As you build each mission, you'll see a new action you can complete on your test track.

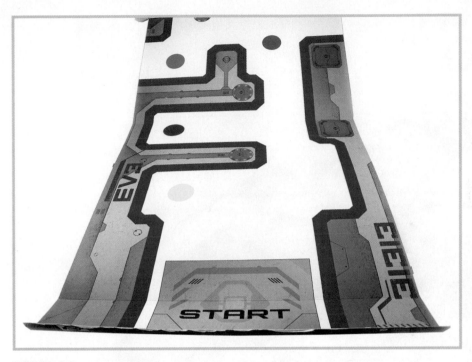

FIGURE 4.3   Use your test track for this mission.

Mission 2

The blades go away for Mission 2, and the Track3r gets a ball shooter, as shown in Figure 4.4. You can place stacked tires on the indicated areas at the end of your test track, and the Track3r will shoot them. It will only shoot them if you put both the tires and the bot in the spots indicated for them on the test track. There is no attempt to sense where things are or compensate for placement differences.

TIP

You should pay attention to how the shooter is constructed in this build because you can reuse the same technique whenever you'd like to build your own shooter.

FIGURE 4.4   The Mission 2 robot with ball shooter.

Mission 3

For Mission 3 you take off the ball shooter and add a gripper, as shown in Figure 4.5. If you place the Track3r and tires on the designated spot on the test track, the bot will grab the tires and place them on another spot.

TIP

This is one of many ways to create a gripping arm. You should pay attention to this build and see whether you can think of ways to improve the gripping power in future builds you create.

FIGURE 4.5   The gripping style of Mission 3.

Again, the bot makes no attempt to compensate for different conditions, so you must place everything exactly as instructed.

By this point, you should be seeing some real possibilities for your EV3 with just a few variations in engineering and programming. If you can make a tank bot that knocks down items, why not change the "blades" into a broom and make a robot that cleans your floor? If you can program a robot to hit a target, you could tweak the same program to make a robot that avoids those targets.

Sometimes engineering changes necessitate programming changes and vice versa, but not always. As you build a robot, you might go through several iterations to find the most efficient design.

NOTE

One of the best things you can do with these instructions is modify them. One enterprising twelve year old, Shubham Banerjee, modified one of the user-submitted bonus models, the Banner Print3r, to create a low-cost braille printer for the blind.

Mission 4

Figure 4.6 shows the end result of Mission 4. At this stage, you can get rid of the test track and use your bot on a flat surface. A hammer replaces the gripper. This Track3r variation uses the infrared sensor to detect objects, turn around, and try to crush them with the hammer. Not only is this mission a great example of programming with the infrared sensor, it shows you just how versatile the medium motor is. It has now powered a ball shooter, a spinning blade, a gripper, and the hammer.

FIGURE 4.6 The Mission 4 robot gets fancy with its hammer.

Mission 5

Mission 5 brings the Track3r full circle and it gets the whirling blades back, as shown in Figure 4.7, but this mission also adds in the use of the remote control. You now control where the tank goes and whether the blades spin. You could use the Track3r on the test track or on any flat surface.

You might be building to see how the engineering works this time around, but pay attention to the differences in coding, too.

FIGURE 4.7 The Mission 5 robot adds in some whirling blades.

R3ptar

The R3ptar, a robotic snake, is one of my favorite core builds. There are only two building stages for this mission, so it's relatively fast.

Three programs come with the R3ptar instructions:

■ Program 1 plays rattling sound effects and moves the bot. It is meant to test your connections on Mission 1, and it's the only program you can run with Mission 1.

- Program 2 uses the infrared sensor to detect and strike at objects near the snake-bot.
- Program 3 enables you to use the remote control to manually control noise and movement.

Mission 1

Mission 1, as shown in Figure 4.8, builds the snake without a head. This is a good time to test your bot, just to make sure you've plugged in everything correctly. Run Program 1 and double-check that the sound effects play and that the robot moves a bit like a snake. If nothing happens, you know you have to go back and troubleshoot your connections and build.

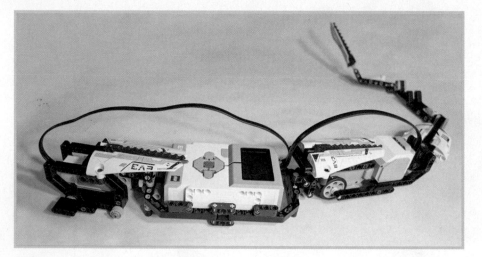

FIGURE 4.8 The headless R3ptar.

Mission 2

Mission 2 completes the bot's look with a head that uses the infrared sensor as eyes and the spiked decorative bushings as teeth (see Figure 4.9).

Make sure you put your robot on a surface with plenty of room, and be sure that you only try Program 2 with people who will appreciate the surprise of having a robot snake strike at them.

CAUTION

Testing this bot with a pet is not only cruel, it could result in damage to your EV3 if an animal becomes surprised and aggressive.

When you run Program 3, you are relying completely on the beacon/remote, so it will no longer strike at anyone unless you press the button to make it happen.

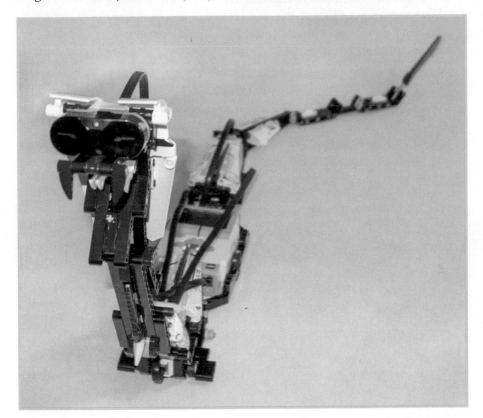

FIGURE 4.9 The completed R3ptor.

Spik3r

Spik3r is a spikey spider or scorpion. The bot works best on large, clean floors, because it moves around and shoots balls. You build this complex bot over five missions, so budget extra time for this one.

Mission 1

Mission 1, shown in Figure 4.10, is just to build and shoot the ball-shooting scorpion tail. This gives you the chance to troubleshoot the shooter before you complete the rest of the build. That's probably a good idea whenever you're building any project, but these demo robots make the concept more concrete.

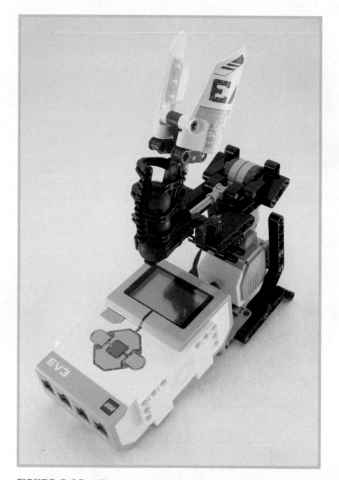

FIGURE 4.10   The completed first mission.

Mission 2

In Mission 2, you add six legs to the bot, as shown in Figure 4.11. The programming mission tests the leg movement along with the ball shooting.

Check out the way the leg build works at this point. There are six legs and only two large motors, so the leg motion has to be divided up in a way that still looks mostly like insect movements and yet allows for the same motor to control multiple legs.

FIGURE 4.11 Mission 2 adds the legs.

Mission 3

In Mission 3, you add pincers to the front of the Spik3r, shown in Figure 4.12. The program uses those pincers to "attack," although at this point the attack is not intelligently guided.

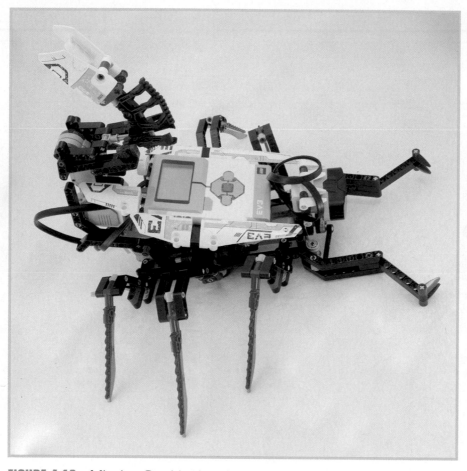

FIGURE 4.12   Mission 3 adds the pincers.

Mission 4

Mission 4 adds infrared sensor control, which searches for objects to attack with the pincers and ball shooting tail. This is an independent action. Figure 4.13 shows the build. Eventually it will search for and attack the beacon.

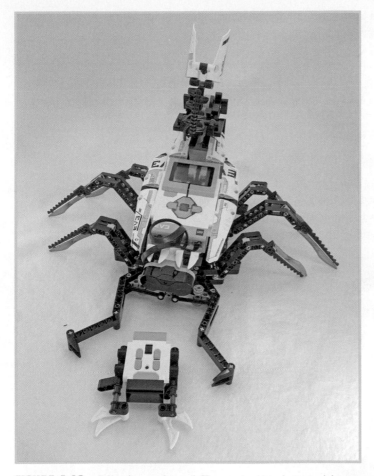

FIGURE 4.13   Missions 4 and 5 are both pictured here.

Mission 5

Mission 5 adds a cute "bug" created from the remote control, which is also shown in Figure 4.13. The remote in the accompanying programming mission is used as a beacon, so the bot will search for the bug beacon and then shoot it with balls and attack it with pincers. This is the robot all of those missions were working to create.

Ev3rstorm

Ev3rstorm is a punk-rock skating bot built in six missions. Rather than using the usual tank driving motion for treads, this bot has legs that skate along on those treads. This bot is also the most humanoid of the core builds and features prominently on the cover of the EV3 box. You'll use most of the beams with this set, so make sure you don't lose any parts before you start and set aside several hours of building time.

There are six missions in the Ev3rstorm build, which indicates a long build.

Mission 1

Mission 1, as shown in Figure 4.14, is just putting together the legs and tank treads. The program tests the gliding and skating motion in a partial figure-eight pattern. This lets you know if you've hooked everything up correctly.

FIGURE 4.14   The skates and not much else.

Mission 2

In Mission 2, your bot starts to look more humanoid (see Figure 4.15) because of the addition of arms and a pincer. The program uses the touch sensor to activate an abbreviated skating pattern. It's no longer the figure eight pattern from the first mission. The arms and pincer are just there for show. You would need an extra motor to make those parts move as well.

FIGURE 4.15 The build starts to look more humanoid.

Mission 3

In Mission 3, you add a blade hand, as shown in Figure 4.16, and this mission enables the use of both the touch sensor and the color sensor to control the bot's motions. The remote will be used later, but this bot does not yet have the infrared sensor in place.

FIGURE 4.16 This robot skates and can be controlled.

Mission 4

Mission 4 adds the infrared sensor as an additional set of eyes (see Figure 4.17). The program uses the infrared sensor, along with the touch or color sensor, to sense and avoid objects in front of it, so you can wave your hand in front of the bot to change its direction.

FIGURE 4.17 Now your robot is mostly assembled.

Mission 5

Now, on Mission 5, you swap out the blade hand for a ball shooter and attempt to shoot targets. I suggest lining up plastic dinosaurs or other toys to see whether your Ev3rstorm can sense and shoot them.

Mission 6

Mission 6, shown in Figure 4.18, is the same build as in Mission 5 but you add a decorated remote beacon, like you did in Mission 5 of Spik3r. The Ev3rstorm will attempt to find and shoot the infrared beacon.

FIGURE 4.18   The robot is the same for Missions 5 and 6. The only difference is the beacon.

Many EV3 fans find this build both incredibly cool and a little frustrating, as it involves building a lot of parts that were swapped out or removed across each of the missions.

Gripp3r

As its name suggests, Gripp3r is a gripping robot. It is humanoid looking with a spikey head and infrared sensor eyes. There's also a slight problem with the build, in that the plastic wing catches on the treads when Gripp3r has lifted an object, meaning that it makes a ratcheting, clicking noise. Fortunately, this is an issue you can safely ignore, as it doesn't seem to damage the robot.

Mission 1

Like the other builds' first missions, you build and test a single part—in this case, the grip handle (see Figure 4.19).

TIP

This is actually great practice for when you want to engineer more complicated parts. Start with the item you think will be most difficult to build. Test it to make sure everything moves well, and then go on to build the rest of the robot around it.

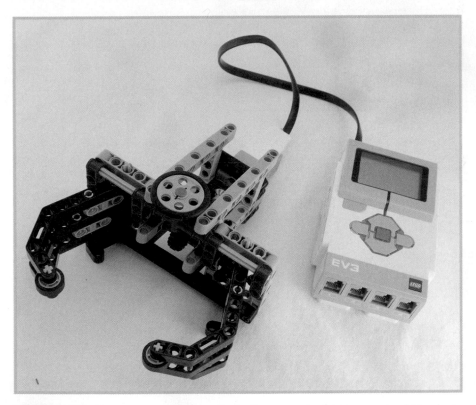

FIGURE 4.19   Test your grip handle before you build the rest of the bot.

Mission 2

In Mission 2, shown in Figure 4.20, you combine the grip you made in Mission 1 with tank treads and add a stacked tire target object for the Gripp3r to grasp.

FIGURE 4.20 Here is both the grip handle and the object it will grip.

Mission 3

In Mission 3 you basically complete the Gripp3r build with an infrared sensor and spiked hairdo (see Figure 4.21).

FIGURE 4.21 In this case, the Intelligent Brick actually faces the back of the robot.

Mission 4

Mission 4 adds in remote control (see Figure 4.22). Try having the bot pick up tomato cans or other objects to see how the grip and lift action work.

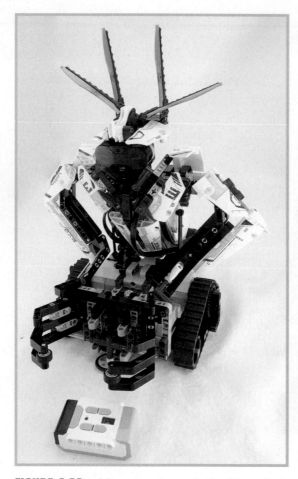

FIGURE 4.22   After you have built this mission, you can control it by infrared remote.

Bonus Bots

MINDSTORMS beta testers and power users have also created plenty of bonus building instructions. These instructions are available at http://www.lego.com/en-us/mindstorms/products/ev3/31313. These are user submissions. You can download them with your browser or use the More Robots button in the lobby of the EV3 home edition software. I cover this in more detail in Chapter 12, "Extending Play."

Here's a list of the bonus builds available at the time I wrote this book. The list and pictures are available in the appendix in the back of this book.

Banner Print3r

Bobb3e

Dinor3x

El3ctric Guitar

Ev304

Ev3game

Ev3meg

Kraz3

MrB3am

Rac3truck

Robodoz3r

Wack3m

Summary

In this chapter, you learned about the basic models for the LEGO EV3 home edition. Going through the models one at a time to see the end result and get experience building is informative and gives you inspiration for creating and engineering your own robots. The emphasis on missions in these builds sometimes is for testing, but missions often show how quickly and easily you can change a bot's capabilities. The next chapter provides a look at the LEGO Education models as well.

5

Building the LEGO Education Bots

Chapter 4, "Building Your First Bots," covered the EV3 home version models and their "missions." If you purchased the LEGO Education version of the EV3, models and instructions are available for you, too. As this chapter covers, the models are different, just as the kits are slightly different, but they're still very exciting builds. The LEGO Education software is sold separately, and no tablet version of the building instructions exists, at least as of this writing.

If you buy the LEGO Education version of the software (an extra $99), you can get the instructions and programs for models that can be built with the basic Education set. LEGO Education also sells additional modular downloads for instructions, programs, and lesson plans. If you're a teacher or team leader, these might be great options. If you're just an individual or hobby builder, you might want the Education and LEGO Education expansion set just for the possibilities. You can get the instructions for all the Education set models from Robot Square at http://robotsquare. com/2013/10/01/education-ev3-45544-instruction/.

> **TIP**
>
> The EV3 retail programming software is free and compatible with the LEGO Education version of the EV3. You need to download additional programming blocks for the gyro and sonic sensor at http://www.lego.com/en-us/mindstorms/downloads/software/ ddsoftwaredownload/.

Educator Vehicle

Figure 5.1 shows the Educator Vehicle, which is just a basic car bot with a few add-ons. Figure 5.1 shows the vehicle expanded to include a lifting arm, gyro sensor, color sensor, sonic sensor, and colorful box to either lift or to launch commands by holding the colors up to the color sensor. All the instructions for building this model are available in the printed manual that comes with the set.

This robot is extremely basic by design. The intent of the Educator Vehicle is to give you a very basic building block for problem solving and a good introduction into programming.

In Chapter 6, "Hacking What You Have," you'll build your own version of the Educator Vehicle using the EV3 home edition kit, and in Chapter 7, "Make Your First EV3 Program," you'll start to program it.

FIGURE 5.1 Here is the Educator Vehicle with the lifting arm installed.

The Educator Vehicle is a great starter bot, because you can program it to do a lot of different tasks. I cover Educator Vehicle programming in more detail with tasks and how they can be programmed in Chapter 7, but Figure 5.2 shows an example of a program to make the Educator Vehicle drive forward until it crosses a black finish line.

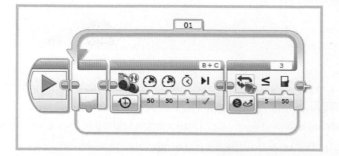

FIGURE 5.2 A very basic loop to make the vehicle stop on a black line.

As far as the engineering goes, the significant part of this bot is really the Technic pivot ball. Chapter 2, "What's in the LEGO Education Box?" explored the LEGO Education set parts, and one of the parts was a ball bearing that fits inside a LEGO socket. This part (see Figure 5.3) allows for easy, frictionless gliding, and it's used in the Educator Vehicle to create a stable third wheel for a lightweight bot that turns easily and moves quickly on flat surfaces. In Chapter 6, you'll try to work around the lack of pivot ball by creating a pivot wheel, but it doesn't work as well as the pivot/castor wheel. If you purchased the EV3 retail set, you can still order this part from LEGO Education at two for $15.

FIGURE 5.3   Castor ball or Technic pivot ball.

Gyro Boy

Gyro Boy is my favorite out of all the basic builds—retail or Education set. Not only is it a cute humanoid, the ability to balance and move on two wheels is very impressive. This humanoid bot, shown in Figure 5.4, uses the gyro sensor to balance on two large wheels. The screen displays eyes that sleep while the bot balances and calibrates its sensor. The eyes open when Gyro Boy walks, and if the bot falls over, the eyes turn into comic-style Xs.

FIGURE 5.4   A sleeping Gyro Boy bot.

The Gyro Boy uses one of every single sensor in the LEGO Education set. The gyro sensor is used for balance. One touch sensor (the Education set comes with two) is used in the back as a reset button for the program. The ultrasonic sensor is on one arm and prevents Gyro Boy from running into objects or walls. The other arm has a color sensor, which is actually used to launch commands by color.

TIP

As your builds get more complex, you might want to label your connecting cables with some tape on both ends. Electrical tape comes in many colors and serves as an easy way to color code cables. This helps you connect your cable into the correct ports.

What you can't see in Figure 5.4 is the small platform you can have Gyro Boy rest upon. There's a touch sensor on the back of the bot that rests on the top of the platform. Figure 5.5 shows the balancing platform, which is also used as a control device for Gyro Boy. You place Gyro Boy on the platform and launch the program. Gyro Boy's eyes close to indicate sleeping. After a short wait, Gyro Boy rolls off of the platform on his own and balances on two wheels like a Segway. After the bot is balancing on his own, you can hold the balancing platform in your hand and control the direction and speed of Gyro Boy by showing different colors to the color sensor. Be sure to not touch the sensor with the color—just put it in front of the sensor where it is in view.

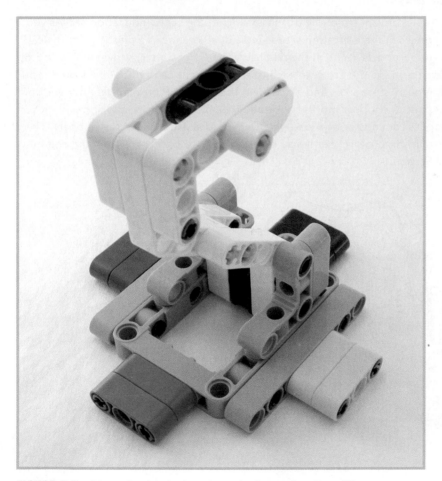

FIGURE 5.5   Here is the balancing platform for Gyro Boy.

NOTE

Gyro Boy is particularly sensitive to problems with the gyro sensor, low batteries, and non-flat surfaces. Always make sure the gyro sensor is calibrated, the batteries are fully charged, and your eye is on Gyro Boy to ensure he doesn't fall off of a table. Falls from that height can break the plastic LEGO pieces or the Intelligent Brick.

Color Sorter

The Color Sorter takes advantage of the track elements in the LEGO Education set. As you might recall, the tracks in the Education set are individual interlocking plastic parts instead of the fixed-size rubber treads found in the retail set. The Color Sorter, shown in Figure 5.6, uses the color sensor on the right to scan each of four different colored double beams. You scan the beams, grocery store style, and place them in the hopper on the left. After you've loaded all the beams, the hopper moves back and forth along the track and sorts all the beams into four piles (or cups or beakers—whatever you place in front of the hopper). The task is not particularly impressive, but it's a great way to learn basic programming and how to use the Color Sensor.

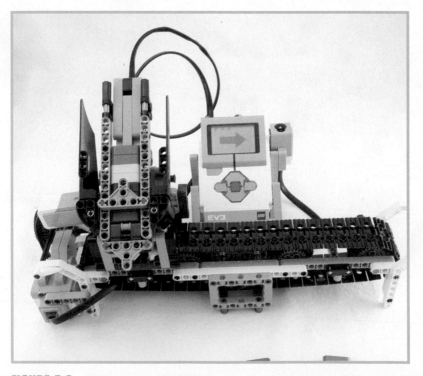

FIGURE 5.6   Scan the colored pieces and then place them in the hopper for sorting.

Puppy

The Puppy bot, shown in Figure 5.7, is a fun model to build with younger kids, although it's still pretty fun for grownups. The puppy can sit, stand, and raise one back leg (with a honking noise). It doesn't actually walk anywhere. If you press on its back, the puppy plays a panting sound.

The build also has a bone and uses the color sensor (hidden underneath the Intelligent Brick "chin") to detect when it is offered and, when it is, play chomping sounds. The Puppy appears to get excited when it is played with and falls asleep when neglected by changing the shape of the eyes displayed on the screen. There's even a special "puppy love" eye display.

FIGURE 5.7 The Puppy posing with its bone.

Pay attention to how the screen and audio enhance the life-like feeling of this build. You might be able to capture that feeling in robots you build.

Arm

The final basic build with the EV3 Education core set is the Arm, shown in Figure 5.8. This simple grabbing arm moves objects a limited, pre-set distance. The engineering for the arm is the important aspect, because it uses a complex set of gears. The grabbing itself is handled by a medium servo, with the lifting and turning done by the large servos.

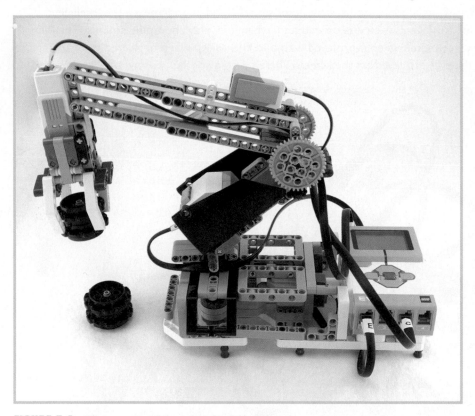

FIGURE 5.8 The arm with a load of tires.

Expansion Models

Chapter 3, "Comparing the EV3 and NXT," discussed the Education version of the EV3 and the EV3 expansion set. The expansion offers a lot of parts and pieces not included with the core set. There are also expanded models that do some fantastic things, such as a walking elephant and a robot that makes spinning tops. The expansion models all call for the LEGO Education expansion set as well as at least one LEGO Education EV3 set. In some cases, you'll need more than one set to complete these builds. The activities are designed for classrooms and teams.

TIP

There's no reason the LEGO Education expansion set is just for the LEGO Education kit. The parts and pieces are compatible with the EV3 home edition. If you find yourself needing more frame beams or different sizes of wheels, consider purchasing the expansion set.

As with the core models, the instructions to build them are available at Robot Square, but programs for these bots are only available if you own the LEGO Education version of the software (sold for $99).

The Elephant

The Elephant, shown in Figure 5.9, is my favorite expansion model. I'm always a fan of a good animal-shaped bot. This build is fairly complex and takes just about every single beam available in the core Education set plus an expansion set. The elephant, once assembled, can walk, trumpet, and use the trunk to pick up a barbell. To do all that, however, you have to command each step by touching the appropriate button on the Intelligent Brick. Unlike the Puppy, it does not attempt to appear autonomous.

FIGURE 5.9 The elephant with an item for lifting with its trunk.

This build is another one you should study for its use of gears. Even without power, you can bend the trunk forward and backward to see how the gears work together to give the trunk its motion.

Tank Bot

The Tank Bot takes advantage of the cleats that come with the expansion set to build a tank with some serious traction (see Figure 5.10). It can climb over objects and would work well in an obstacle course challenge. This is not a remote-controlled bot by default, but you could certainly modify the design with a separately purchased infrared sensor to turn it into one.

FIGURE 5.10 The Tank Bot. Replace the gyroscope with an infrared sensor to make it into a remote-controlled robot.

Znap

The Znap, shown in Figure 5.11, is the LEGO Education equivalent of the Rapt3r build from the retail kit (refer to Chapter 4). It uses the sonic sensor to detect a person or object in front of it and "snap" at the hand or object. The plans, as written, call for simple tank tracks, but there's no reason not to add the red tank cleats to this design for a little extra climbing ability.

FIGURE 5.11   Znap with added red cleats.

Remote Control

The Remote Control build, shown in Figure 5.12, is designed to work with some other build, such as the Znap. Obviously, that means you need two or more EV3 sets to make this project worth building. You can hold the remote in your hand and squeeze the lever at the end, turn the tire knob, and press the touch sensor button to control the movement and snapping action for the Znap bot.

You need two EV3s for this remote to work. Technically, you could build this and many of the other designs by using one home edition and one LEGO Education EV3 kit. However, if you've already got an EV3 home edition, you could just use the infrared sensor and beacon remote that comes with the kit to modify your robot and allow it to be remote controlled.

FIGURE 5.12   The LEGO Education version of a remote control.

The Stair Climber

The Stair Climber does just as it sounds—it climbs stairs (see Figure 5.13). The bot has what can only be described as a wheel elevator. When it runs into a stair, the tracks lift the first four wheels upward, and then raises the rear wheels after the bot has successfully climbed a stair. It can also climb curbs, although watch out for cars when you try that maneuver.

FIGURE 5.13 The Stair Climber backs into stairs and climbs them.

Color Spinner Factory

This gigantic build uses two Intelligent Bricks chained together to find, grab, and assemble color-coded top pieces into a custom spinning top, which it then spins and launches.

This is also the most complex build out of the expansion builds. It requires two LEGO Education sets and an expansion set. I don't own two LEGO Education sets, but I'm happy to say that you can still build something very near the Color Spinner Factory from one LEGO Education set, one LEGO retail set, one expansion set, and plenty of spare parts scavenged from an old NXT and a few assorted Technic sets. Figure 5.14 shows my result.

The biggest difference between my final version and the original plan is that I ran out of colored size 3 beams and friction snaps with bushing ends. The only area where the actual color matters is along the bottom row where the second color sensor is installed. Make sure the correct colors are used along that row, and then don't worry about color matching on other areas of the build.

FIGURE 5.14 My approximation of the Color Spinner Factory build.

The Color Spinner Factory is the sort of project for which you would want a group of eager students and a longer span of classroom time to complete. However, this build is a fantastic demonstration of how you can chain EV3 Intelligent Bricks together to do tasks that are beyond the capabilities of a single brick. Figure 5.15 shows the fully used set of ports on the top of the bot, including the USB port connecting the bots together.

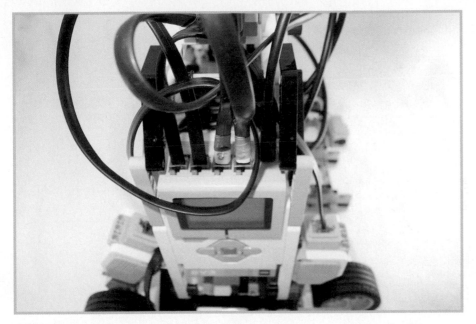

FIGURE 5.15   Notice how the cords are labeled to avoid confusion.

MISSING PARTS

If you have a LEGO Education set, the part you're probably most likely to miss from the retail set is the infrared sensor and remote. You can purchase it from LEGO Education for $29, but the beacon is an additional $29. Ouch. If you own the retail set, the sonic and gyro sensors are $29 each. The expansion set is an additional $99.

Summary

The LEGO Education version of the EV3 has builds available for both the core set and the expansion set. Although you can find instructions for building the models online, the demo programs are only available with the $99 LEGO Education version of the programming software. If you want to program the bots yourself, the education version of the EV3 is perfectly compatible with the retail version of the software.

6

Hacking What You Have

You've now explored all the parts in the box and looked at all the types of robot builds provided in the official LEGO EV3 instructions. It's time to build your own bot. The first bot I recommend trying to build is a car or tank. It is the most common build, and all you need to get started is the Intelligent Brick and two motors. Leave room to attach a light sensor. You'll need that for your first programming lesson, which will start in Chapter 7, "Make Your First EV3 Program."

> **NOTE**
>
> Theoretically, you could make a robot with only one motor, but steering it would be difficult. With only one motor, the robot would have to have the wheels chained together to go only forward or backward or use only one wheel and travel entirely in circles.

This chapter gets you started on building, but it is also builds on skills you learned by building the demo robots from Chapter 4, "Building Your First Bots," and Chapter 5, "Building the LEGO Education Bots." If you haven't built those robots yet, I recommend at least glancing through Chapters 4 and 5 to familiarize yourself with the models. Later on in this chapter, we'll modify those builds.

Project 1: The Car

If you're ready for a challenge: As soon as you finish reading this paragraph, put down the book and build a car or tank with your robotics set. You can make it as fancy or simple as you want, but keep in mind you want to expand it later, so leave some ability to attach more sensors. To make life easier, connect your servos to ports B and C. This is an easy testing shortcut. The "demo" program ships with all EV3 MINDSTORMS, and it uses the B and C ports to control the wheels. Don't worry about spending a long time on this. Just build an initial, working model.

If you're not quite up to this challenge yet, don't worry. Just read along, and I'll show you what I did and how we can make it better.

If you tried the challenge, how did you do? In the following sections I show you just one example car that I built very quickly, without any reference to the EV3 demo projects. I built this car using the LEGO Education set, but you could just as easily use the home edition set.

Here are the steps to create this simple car build:

1. Place the treads on two small wheels and gather an axle for the front wheels (see Figure 6.1).

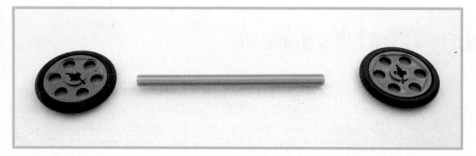

FIGURE 6.1   The wheels and axle.

As you can see, the wheels have axle holders. That means you can use servos to move the wheels. In this case, you want the wheels to freely spin.

2. Attach the wheels to the beam frame using an axle (see Figure 6.2). Using a single axle allows the wheels to spin in unison while being held securely.

3. Place two blue long (3M) pins on the side of the beam frame.

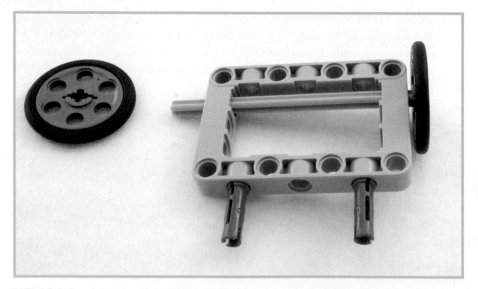

FIGURE 6.2   Attach the wheels and pins.

4. Connect the beam frame and wheels to the bottom of the Intelligent Brick with the blue pins (see Figure 6.3).

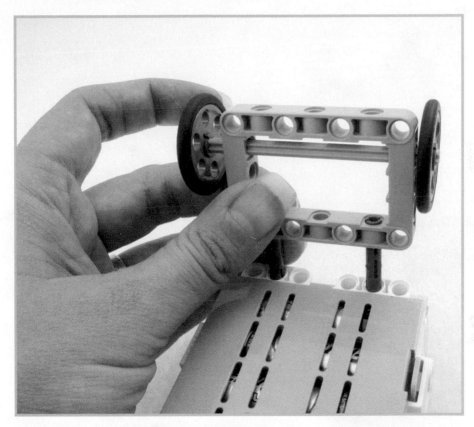

FIGURE 6.3   Connect the wheels to the Intelligent Brick.

5. Gather angled beams and cross blocks, as shown in Figure 6.4. These form the back wheels. Make two, and make sure they mirror each other, because they will go on opposite sides to connect the rear wheels.

FIGURE 6.4 The parts for one rear wheel.

6. Connect the parts together as shown in Figures 6.5 and 6.6.

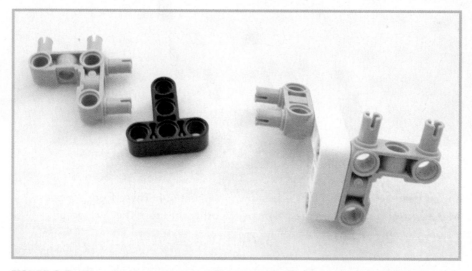

FIGURE 6.5 Remember that you're connecting two mirrored sides.

FIGURE 6.6 The finished support.

7. Now comes the motor power for the rear wheels. Gather the parts shown in Figure 6.7. Connect the rear wheel directly to the large motor using a shorter axle. You'll need two mirrored sides.

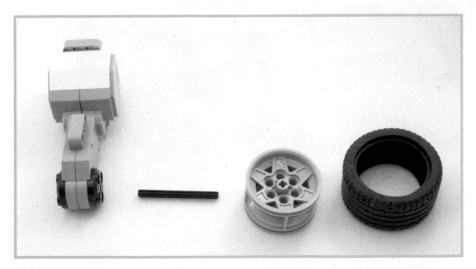

FIGURE 6.7 Gather two sets.

8. Now everything for the car goes together. Fit the connecting beams and cross blocks into the large motor and wheel assembly (see Figure 6.8).

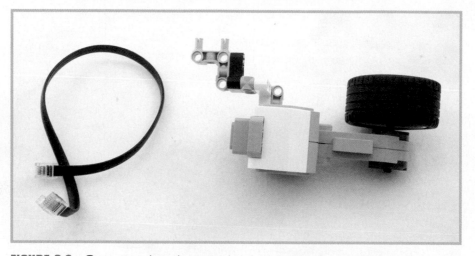

FIGURE 6.8   One completed rear wheel and the connecting cord.

9. Connect the two mirror-image sides to the Intelligent Brick, as shown in Figure 6.9. (I disconnected the motor for the photo to make it easier to see the connection.)

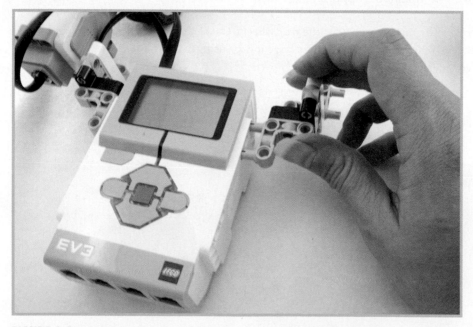

FIGURE 6.9   Attach the rear wheels to the Intelligent Brick.

10. The final stage is to connect the motors to the Intelligent Brick. Plug the connecting cable into ports B and C, crossing the lines.

Any motor port would work just fine to power your motors. The demo program that came pre-installed on your Intelligent Brick uses those ports, so the easiest way to test your car is to use the same ports. After you move on to writing your own programs, you can use whichever port is most convenient.

You can see the final bot in Figure 6.10.

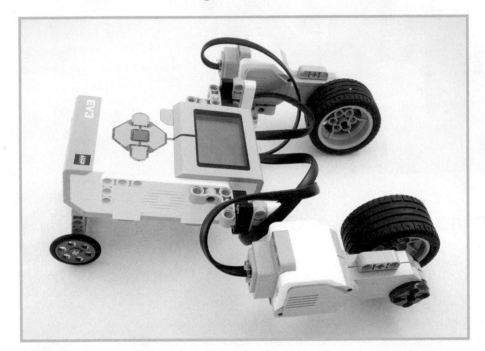

FIGURE 6.10   The finished simple car.

Testing

To test the bot, run the demo program that comes with the Intelligent Brick. Press the center button on the Brick. When the EV3 powers up, the selected tab on the screen should be the left Play tab, as shown in Figure 6.11. The very first program, provided you haven't deleted it, is the "demo program." Press the center button on the Brick again to launch the demo program. You should immediately see eyes and both wheels should turn.

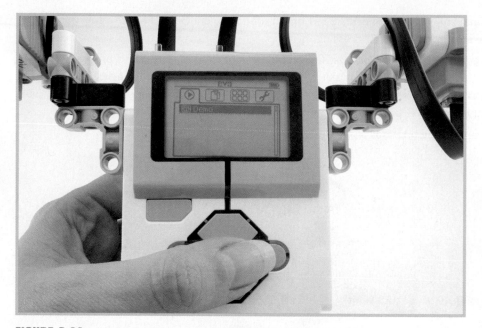

FIGURE 6.11 Launch the demo program.

If you run the demo program again with the bot on a flat surface, it should move and turn. That lets you know the bot works. After you run it on a flat surface, you should also see how it doesn't work (as I describe in the next section).

Troubleshooting the Flaws

You might not be able to see the flaws if you didn't build this model along with the book (and it's perfectly fine if you didn't), but this build has a couple of problems. The car is unstable. When you press the center button, the entire bot bows downward, because the rear wheels are only supported by two pegs on each end, with very little cross support. It cannot bear any serious weight. That's fine if you're just using a small, flat, test track, but delicate designs like this risk snapping pieces and parts.

If you wanted to fix this problem, you could add more cross support or move the wheels closer together.

The car can run on flat areas, but it can't climb obstacles well. That isn't important for this project, but it is something to keep in mind. Additionally, the front wheels are locked together, which limits the turn radius, and there's not much room to pin on extra sensors or parts to expand this vehicle.

You could resolve the climbing issue with tank treads, and you might consider redesigning the front wheels with some sort of differential to make them spin at different rates during turns.

TIP

Failure is fantastic. If you don't fail several times when you are building something, it means you are not challenging yourself. Keep trying new designs and schemes to build improved robots.

Project 2: Hack Your Tank

How do you improve on the car build? One fantastic way to start is to not start from scratch, but to hack what you have. Take one of the pre-existing models and modify it to meet your needs. The most obvious choice for the home edition EV3 set is the Track3r, because it already works perfectly with the demo program and has room to expand with extra sensors. Let's break down the steps for the basic tank:

1. Build up to page 12 of the home edition EV3's printed instructions for the Track3r, as shown in Figure 6.12. You don't need to bother with the decorative wings, small motor, or sensors. Without them you still have a very solidly built tank that will work well for your purposes.

FIGURE 6.12 Page 12 of the Track3r.

2. Add a color sensor, pointing as close to straight down as you can manage. One way to do that is to use an angled beam, a long blue peg, and a short axle (see Figure 6.13). Use another angled beam on the other side.

You're adding a color sensor for your first programming project, which detects when your robot has crossed a black finish line. The sensor needs to aim down because it needs to detect the line on the floor, and it should be fairly close to the floor—half an inch or less ideally—in order to detect reflected light bouncing back to the sensor.

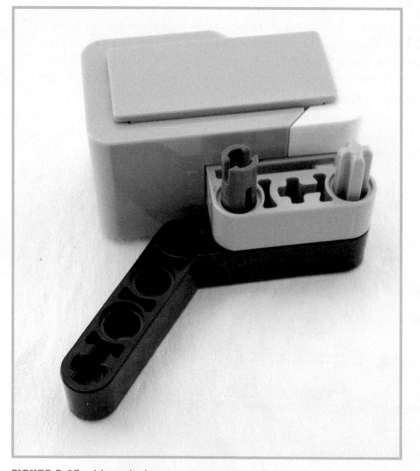

FIGURE 6.13 Here is how our sensor will be connected to the tank body.

3. Make the supports to pin the sensor onto the front of the tank. Two black pegs go on the outside of each angled beam. Mirror axle connectors and the axle to pin connectors to angle the sensor as shown in Figure 6.14 and Figure 6.15.

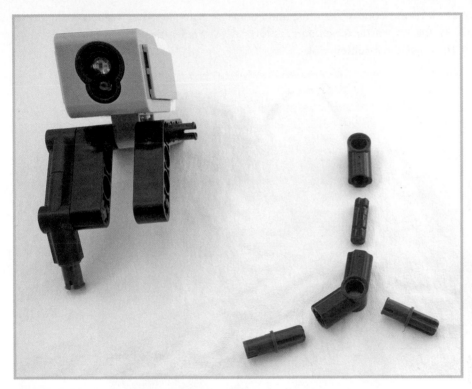

FIGURE 6.14   The color sensor attachment.

FIGURE 6.15   A side angle to let you see how this is built.

4. Pin the sensor to the front of the tank, and connect it to one of the numbered sensor ports. It doesn't matter which port, as long as you remember which one you used. Figure 6.16 shows the resulting tank.

FIGURE 6.16 The sensor attached to the modified Track3r.

This tank is nice and stable. It can actually handle some uneven surfaces, and it will not bow when you press on it. The undercarriage is also well supported with frame beams (see Figure 6.17).

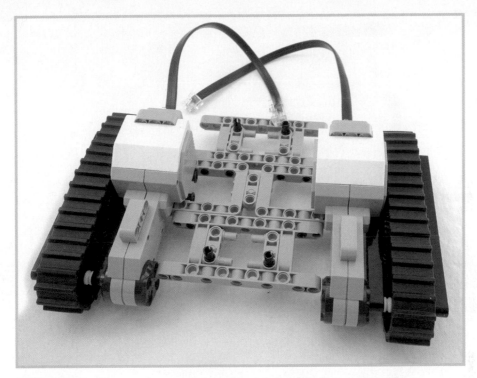

FIGURE 6.17   The undercarriage without the Intelligent Brick.

Project 3: Modify the Educator Bot

The drawback to using the tank is that it is very wide, and it will still make relatively wide turns. What if you want something a bit more nimble? One solution is to use the LEGO Education set's Robot Educator, which is shown in Figure 6.18.

FIGURE 6.18　The Robot Educator.

This nimble bot is designed to accommodate various sensors so you can learn about programming. If you have a LEGO Education set, you can just use the printed instructions to build it.

If you don't have a LEGO Education set, the Robot Educator has the ball caster part, which isn't found in the home edition of the EV3. You could buy a separate ball caster; it's a useful part, but you can't just go buying parts every time you want to build something. Fortunately, you can improvise and hack using what you already have.

Rather than a ball caster, for this project you'll make a wheel caster similar to what you find on the bottom of office chairs and shopping carts. You'll also have to make a few more adjustments for differences in the size of wheels and length of axles. It isn't quite as maneuverable as a smooth ball bearing, but it will work reasonably well for this project.

Figure 6.19 shows both the LEGO Educator build and the hacked EV3 home edition version. The home edition version has the light sensor attached.

EV3 Home edition

EV3 Educator

FIGURE 6.19   Comparing the Robot Educator and a modified home edition.

The following sections take you through the breakdown for this build.

Modify the Design

The first step is to build most of the LEGO Educator vehicle. You'll find the instructions at Robot Square: http://robotsquare.com/2013/10/01/education-ev3-45544-instruction/.

You'll need to make some adjustments from these instructions, however. The home edition EV3 set doesn't have the 4M axles called for in Steps 8 and 20, so you'll need to use the 5M size, which means the axles will stick out 1M past the sides of your build (see Figure 6.20).

FIGURE 6.20 The axles stick out just a little.

The other difference you should notice is that the wheels on the home edition set are slightly smaller than the LEGO Education wheels. That necessitates a second adjustment on the front of the vehicle (see the following section). The LEGO Education version sits up a little higher, so you can adjust the angled beams used on the front of the vehicle in Steps 8 and 20 to sit slightly higher. The beams will be black instead of white, and they're also a little longer, but that's okay. Figure 6.21 shows the modified version with longer, black angled beams in front.

FIGURE 6.21   The modified robot front.

Substitute for the Caster Ball

For the next adjustment, you need to substitute something for the caster ball used on the LEGO Education version of the bot. My first thought was to trap one of the red balls that come with the EV3 in a box of beams and pegs, but the sizes didn't match up. Instead, let's make something more like the swivel wheels on shopping carts. In other words, you can substitute a caster *wheel* for the caster ball.

There's a second problem: height. Whatever solution you make has to keep a fairly low profile, because the rear wheels on the EV3 home edition set are smaller than the wheels on the LEGO Education set.

One solution is to use the smallest wheels and make a sort of peg-pivot to connect it to the top, as follows:

1. Gather the parts shown in Figure 6.22.

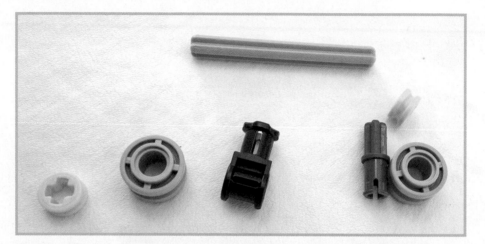

FIGURE 6.22   The parts for our castor wheel.

2. Put the axle through the cross hole connector, as shown in Figure 6.23.

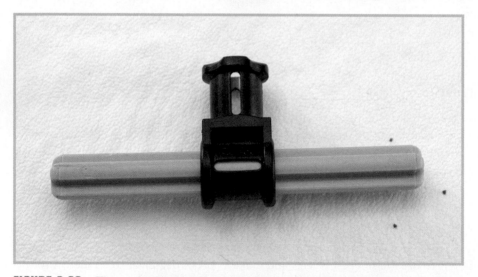

FIGURE 6.23   The axle placed through the cross hole connector.

3. Place the wheels over the axles. One advantage of the smallest wheels is that, unlike the other wheels, they have round openings, which means that they'll spin freely.

4. It also means they'll slip right off of this axle, so put half-bushings on either side of the wheels to keep them on the axle. You should now have something that looks like Figure 6.24.

FIGURE 6.24   The wheels and bushings in place.

5. Put on the blue half-axle connector, as shown in Figure 6.25.

FIGURE 6.25   A blue friction pin to add some spin.

There's another compromise you're making by using this blue piece. This piece has friction, meaning it will actually slow down the twisting motion. Sometimes that's what you want, but in this case, it's not. The wheel still turns, but not as freely as other pieces. How important is this? Build the bot and experiment.

NOTE

The reason we don't put the treads on the front tires is because it provides too much friction when making turns. The bot will jerk through turns rather than glide.

Now you need to attach the castor wheel to the front of the robot.

6. There are no centered bottom-facing peg holes on this bot. You have to make one, so grab the two cross blocks shown in Figure 6.26.

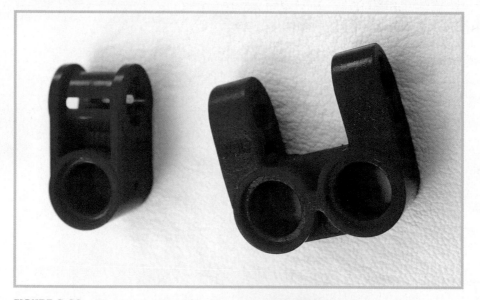

FIGURE 6.26 These parts will make our bottom-facing peg hole.

7. Connect them to the bottom of the bot using an axle, as shown in Figure 6.27.

FIGURE 6.27   The axle turns the parts into bottom-facing peg holes.

8. Place the finished wheel in the center spot, as shown in Figure 6.28.

FIGURE 6.28 Put the castor wheels on the robot.

That should be it. On my bot, this configuration was adequate but not optimal. Test your bot. If it does not turn smoothly like a good caster wheel, you can try using a peg connector to add frictionless spinning.

Be aware that if you do this, you're going to have to connect it higher on the bot to avoid tilting the frame. You also have to worry about stability. Can your bot support itself on the front wheel? Check to see if you have some spare Technic parts from other LEGO sets you have around the house. It's possible that you may have the makings for an improved design.

Experiment with the design until you have a smooth, stable caster wheel. After you are finished, place and connect the color sensor on the front of the bot so that it points straight down. It should look similar to Figure 6.29.

FIGURE 6.29 The final version with color sensor attached.

It doesn't matter which sensor port (1–4) you use, as long as you remember which port you chose.

Remember, this sensor is going to be used in Chapter 7 to let your robot detect when it has crossed the finish line. It should be facing downward and fairly close to the ground without actually scraping it. The aim is to have reflected light bounce back into the sensor.

Summary

In this chapter, you explored robot building by using and modifying existing plans. Hacking an existing robot build means you can take advantage of others' solutions while still being innovative with your own designs. In the next chapter, you learn how to program these LEGO robots.

Make Your First EV3 Program

This chapter gets you started on EV3 programming. You're going to take the vehicle you built in Chapter 6, "Hacking What You Have," and turn it into a self-directed robot that stops on a line. This basic program should familiarize you with the EV3 programming interface.

If you have experience using the NXT 2.0, you'll be familiar with many of the basics of LEGO's programming system. However, there have been updates with the EV3 system, and programming is now easier and more advanced than in earlier systems. As with any chapter, feel free to skip to the parts that interest you most.

THE MANY USES OF LABVIEW

EV3 programs use a version of LabVIEW developed by National Instruments. LabVIEW is actually used by scientists and engineers in all sorts of real-world projects, such as embedded chips, wind turbines, and even self-driving cars. The two versions of LabVIEW do not look completely the same, but it's still exciting to think that programming your EV3 is giving you potential practice time for programming future self-driving cars.

At this point in the book, you should already have downloaded and installed the EV3 MINDSTORMS home edition software. Some differences exist between the LEGO Education and home editions of the LEGO MINDSTORMS EV3 software, as discussed next, but you may use either version for this exercise.

About the LEGO Education Software

The LEGO Education version of the EV3 software is a $99.95 purchase from LEGO Education, and its tutorial programs and model building instructions are geared toward use with the LEGO Education kit, as discussed in Chapter 5, "Building the LEGO Education Bots." In addition, the LEGO Education version of the software offers two unique features out of the box: data logging and curriculum building.

The data-logging feature enables you to see a graphical representation of sensor data in real time from a connected EV3. You can also manipulate the data and use it for programming bot activities along sensor thresholds. It's a powerful feature designed to showcase science experiments in the classroom.

Teachers and team leaders can use the curriculum building feature to design their own lessons for students. The software is also expandable. LEGO Education sells special curriculum packages for engineering projects or space exploration themes.

These curriculum packages are useful if you are teaching a classroom or leading a homeschool group, but they aren't as necessary for the lone builder with a single home edition of the EV3. Individuals at home don't have the logistics problems of working with groups, and they don't generally need to align specific projects to classroom learning objectives.

This chapter focuses on the EV3 home edition software because it is available to everyone. The download is free, and you can use it with any EV3. Basic programming using the graphical interface is the same on both home and LEGO Education editions. If you're using the LEGO Education version of the software, you should be able to build along without missing a beat.

Getting Started

You (hopefully) have already had a chance to download and install one or more programs onto your EV3 when you built the demo robots from previous chapters. This section covers the basics of transferring programs from your computer to your bot. This material should be review if you've already mastered the skill.

Let's get started.

Navigating from the Lobby

When you launch the EV3 home edition software, a splash screen resembling Figure 7.1 appears. It makes noise as the demo robots move around and entice you to click on them. Let's go over the parts of this screen from the bottom up. This entire screen is called the Lobby.

See and launch projects

Quick Start area

FIGURE 7.1   The Lobby is the opening screen when you launch the EV3 home edition software.

The very bottom of the screen is the Quick Start area, which contains tutorial movies. Use the arrow for navigation. You can also switch tabs to see the News tab with announcements or the More Robots tab for bonus robot instructions and programs.

Moving up the screen you see the robot area, which has been rattling and humming at you as you examined the other areas. This area launches the demo robot projects with instructions and programs. Click on the arrows on the left and right sides to tab through the screens for more robot choices. If you click on a robot, red-and-white target-like dots appear on the robot. Clicking on a dot, as shown in Figure 7.2, reveals details about what that particular part of the robot does. For instance, the shooter ball on the Ev3rstorm can aim and shoot.

EV3RSTORM

V3RSTORM moves round on tracked kates and can attack with his Spinning Tri-lade or Blasting azooka. He can detect bjects, follow the red Beacon and ive commands emotely.

Rim and Shoot
With his Blasting Bazooka, EV3RSTORM can shoot ei upwards or straight, deper distance to the target.

CREATE!

FIGURE 7.2 Use the arrows to view more robots.

If you click the Create! button on the bottom left of the screen, you can launch the building tutorial and load the demo programs associated with the missions. If you click on the Open Recent tab on the upper right area of the Lobby, you can open any projects that you've recently worked on, whether they were demo programs or your original creations.

However, this chapter isn't about building robots and using the demo programs. It's about making your own. You're about to create your very first program from scratch!

TIP

A project is an overall package that contains programs and all the images and other files needed to run them. You can also work on several projects at once, and save and share your progress with others.

Creating a New Project

Before you can start a new program, you need to create a new project. You do this by going to the Lobby and clicking on the plus tab at the top of the screen (see Figure 7.3).

Click on the
plus sign

FIGURE 7.3   Click on the plus to launch a new project.

When you click this symbol, the project area appears. You should see a screen similar to that shown in Figure 7.4.

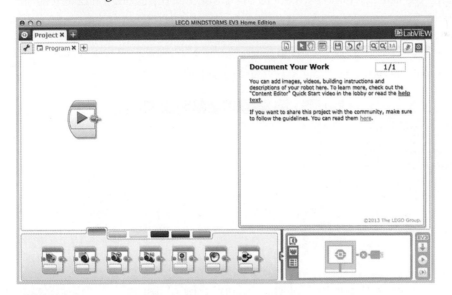

FIGURE 7.4   This is how a new project appears.

Getting to Know the Programming Canvas

You've created a new project, which was pretty simple, right? Now it's time to take your first real steps in learning about the software's interface, or canvas, starting with the top row shown in Figure 7.5.

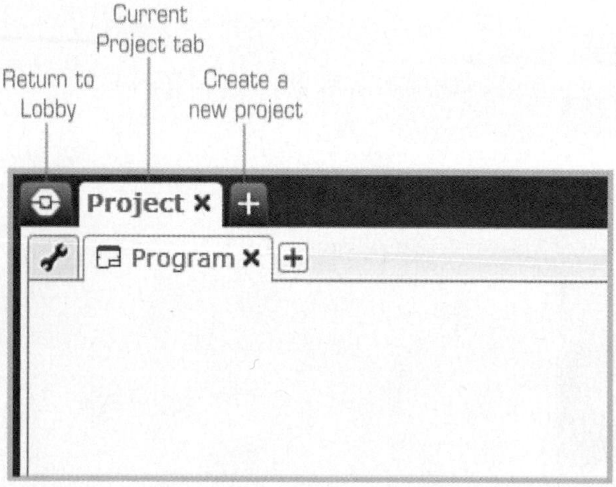

Current
Project tab

Return to   Create a
Lobby   new project

Project ✕ **＋**

🔧 **▢ Program ✕** **⊞**

FIGURE 7.5   This is the top row of your programming canvas.

The biggest change to the top row once you launch a new project is that you now have a Project tab open. The MINDSTORMS symbol to the left of the Project tab still takes you back to the Lobby. The right plus symbol launches another project. You can have several projects open at once, which is handy if you want to copy elements from one project to another or reuse items from one of the demo projects.

Let's look at the left side of the next row, shown in Figure 7.6.

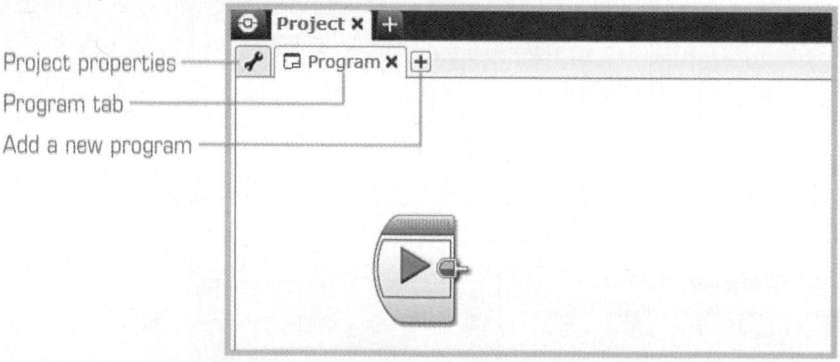

Project properties ──── **Project ✕** **＋**

Program tab ──── 🔧 **▢ Program ✕** **⊞**

Add a new program ────

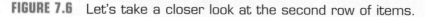

FIGURE 7.6   Let's take a closer look at the second row of items.

The wrench on the left shows you project properties. We'll get to that in just a bit. For a new project, there will be no properties. If you're launching one of the demo programs from the Lobby, you'll see a list of project programs, images, sounds, the description of the project and title, and so on. After you create a project you're proud enough to share, the project properties tab actually gives you the tools to share it with the MINDSTORMS community.

The Program tab (just below the Project tab) is the area where you create your programs. Clicking on it now doesn't do anything, but if you are in the project properties tab or in another Program tab, clicking on the Program tab returns you to the programming area.

The final plus tab adds programs. Don't confuse this with the add project tab. You can have multiple programs within your project, but you want to keep those programs within the same project if they're designed to work together.

The right side of the second row, shown in Figure 7.7, shows a few more commands for the programming area.

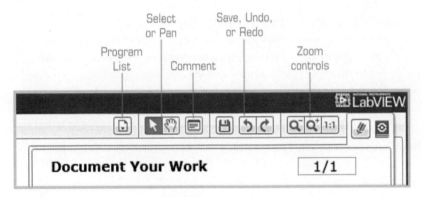

FIGURE 7.7 This is the right side of the canvas toolbar.

If you hover your cursor over a button, its name appears, which should provide you with a hint on its function. You can learn more about these controls, as well as some others, in the following sections.

Program List

The first button, Program List, shows you a list of all the active programs within the project as well as the project properties tab. This is handy if you have a complicated project with a lot of open tabs and need a quick way to navigate between them.

Why would you have multiple programs within a project? Sometimes you may want to experiment with different ways to accomplish the same task or test out small parts of a program before adding it to a larger program. Sometimes you may want to have several different related programs that you wouldn't need to run at the same time, such as a program for use with a remote and a separate program for use without one.

Select or Pan Button

The Select or Pan button (showing an arrow and hand) is how you drag individual programming blocks around on your screen. The hand is the Pan tool, and it enables you to move your view of the programming window around. If you have a very large program and want to work on a specific section, you can use the Pan tool to move your window over to the right area.

Comment

A comment's function is to relay information to programmers and it's an extremely useful tool, both for the original programmer and for anyone who follows-up or modifies what they've created. When you use the Comment button, it makes a little comment box. You drag the box around on your programming canvas next to a bit of program, and start typing a comment in the box.

By itself, a comment doesn't do anything. It's like a sticky note that you can leave for yourself or others to let you know why it is that you made a particular choice. You would be amazed at what you can forget in even a few days, so leaving comments for your future self makes your work go much faster and smoother when you're revising or looking for bugs to fix. Useful comments also make professional coders very happy, so using them is a great habit to develop if you want a professional career as a programmer.

As mentioned, the Comment button is for short notes. If you need longer or more detailed information, you should use the Content Editor, which I show you later in this chapter.

Save, Undo, or Redo

The Save, Undo, and Redo buttons should be familiar if you use a lot of Office-style applications. The Save button sports the old-fashioned disk icon and lets you save your entire project. The arrow curved to the left is the Undo button and lets you undo the last action you took in the program. The arrow curved to the right is the Redo button and lets you redo whatever it was that you mistakenly undid.

Zoom

The Zoom buttons are grouped together to let you manipulate your programming window. The magnifying glasses make the program blocks bigger or smaller, and the 1:1 button returns everything to the default view. You can use these buttons in combination with the Pan button to take a look at exactly the right section of program.

The Content Editor

Finally, notice the last item on the top row, the Content Editor, as shown in Figure 7.8.

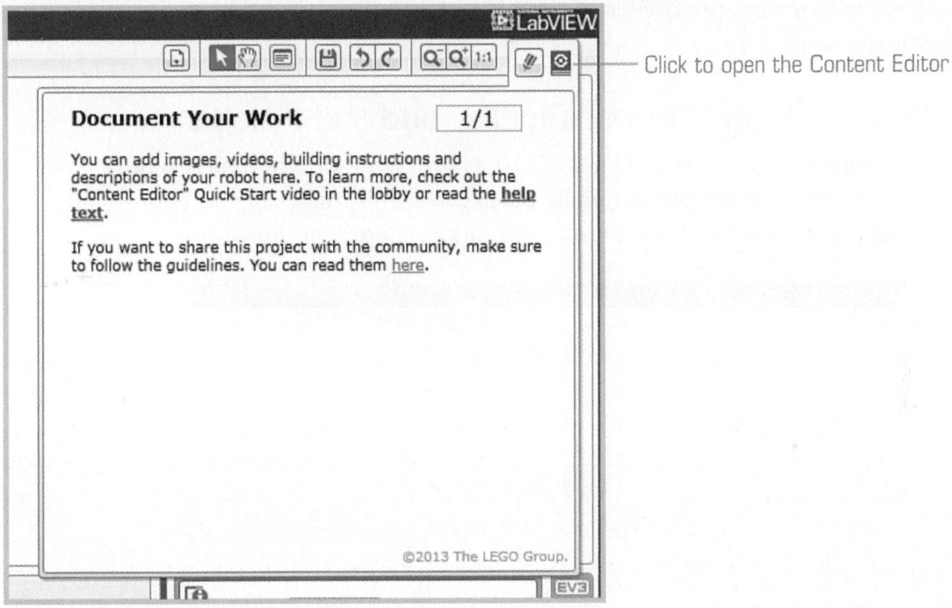

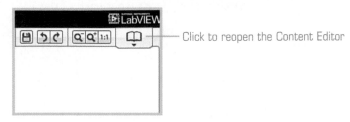

FIGURE 7.8 Use the Content Editor for longer notes and detailed project overviews.

When you launch a new project, by default the Content Editor is open with this screen telling you to document your work. Using this feature is how you can showcase your project with text instructions, pictures, video, links, and interactive elements. You should make heavy use of the Content Editor if you want to make programs to share with the MINDSTORMS community.

Click the Pencil button to start editing your documentation. You do that a little later on in the book, so click the Close Content Editor button (the MINDSTORMS symbol) to the right of the Pencil button. This should close the whole window and get it out of your way, so you can see the programming canvas.

After the window closes, the button should look like a book, as shown in Figure 7.9.

FIGURE 7.9 The Content Editor is closed.

Clicking the book reopens the Content Editor at any time. This way you can document your work as you go.

The Programming Canvas, Blocks, and the Palette

The programming canvas, shown in Figure 7.10, is your main programming area. Immediately beneath it is the programming palette, which is the source for your programming blocks. Each of these little blocks has a different purpose.

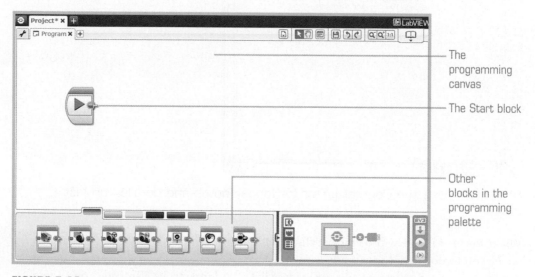

FIGURE 7.10   Here is the programming canvas and palette.

By default you start out with only one block: Start. As you build your program, you'll drag and adjust different programming blocks from the palette below to make your program. You can have loose programming blocks hanging around on your canvas as you build programs, but until you connect them to the Start block, they won't do anything. I discuss how to do that in more detail in this chapter, in the section "Writing Your First Program."

Notice the color-coded tabs at the top of the programming palette. Click on each color to see the programming blocks that fall into that specific area. Here's what each of the colors means:

- **Green is for Action**—These programming blocks make your robot move, make noise, light up, or display something on the screen.
- **Orange is for Flow control**—These blocks can be used to make part or all of the program loop, pause, start, or stop.
- **Yellow is for Sensor**—These blocks are used to measure and sense. You can get feedback from sensors, but you can also get feedback from things you wouldn't ordinarily consider sensors, such as detecting how many rotations a motor has made or how much time has passed.

- **Red is for Data Operations**—These blocks help you do math and control numbers. You can also create variables or generate random numbers.
- **Dark Blue is for Advanced**—Advanced blocks are for doing things such as using the Bluetooth or sensing raw data.
- **Teal is My Blocks**—You can create your own programming blocks. If you have an action you want to repeat in multiple projects, you can make that action into its own block and load it into your programming palette for handy use. For example, you might want to make your name appear on the screen in all your own programs.

The Connection Area

The connection area, which is located next to the programming palette, is the area you'll use to connect your EV3 to your computer and get your program onto your EV3 (see Figure 7.11).

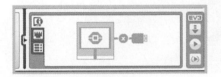

FIGURE 7.11 Here is the connection area.

You can connect your EV3 using a USB cable, or wirelessly using Wi-Fi or Bluetooth pairing. Hopefully you were able to successfully connect and transfer files when you tried out a few demo robots in Chapter 4, "Building the First Bots," and Chapter 5, but don't worry if you haven't. I'll walk through all the steps after we finish our first program.

Writing Your First Program

Now that you've launched the EV3 home edition software and started a new project, it's time to actually start programming your bot. For this exercise, use the LEGO Educator bot you built in Chapter 6, which should look something like Figure 7.12. It should have the light sensor connected to it and aimed at the floor.

NOTE

If you have the EV3 LEGO Education set, use the instructions for the LEGO Educator bot that came with your LEGO Education set. Otherwise, use the modified version covered in Chapter 6.

FIGURE 7.12   Our modified Educator bot uses the EV3 home edition.

What you want this bot to do for the first program is go in a straight line until it comes across a black line, at which point the bot should stop.

Flowcharting

Sometimes, making a flowchart before beginning a programming project can be helpful by clarifying exactly how the program should function. Figure 7.13 shows an example flowchart for the line-detecting program. You don't need a fancy app to make flowcharts. You can just use paper and pencil. However, if you like a software solution, this flowchart was made with Gliffy, which you can find at http://www.gliffy.com/.

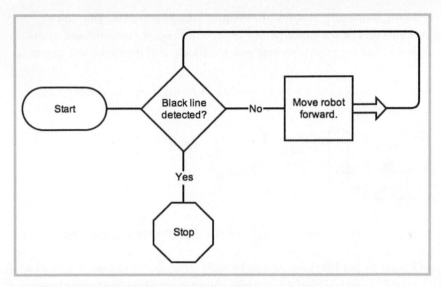

FIGURE 7.13 Here's a basic flowchart for our program.

The line-detecting program just has to go forward when there is no line and stop when the sensor detects the line. As you get further into LEGO programming, you'll see that the EV3 home edition software functions a lot like a flowchart.

Dragging Blocks Onto the Programming Canvas

The programming canvas (refer to Figure 7.11) is where you'll create your program, which you do by dragging programming blocks from the programming palette.

The Start block is already on the canvas when you open a new project, so the next step in the flowchart is to figure out whether your bot senses a black line. This is the "Is line detected?" question from our flowchart in Figure 7.13. Sensing is, of course, what sensors do. In this case the color sensor is attached to the front of the EV3. To find the correct block, follow these steps:

1. Click the yellow tab on the programming palette, as shown in Figure 7.13.

The Sensor programming group

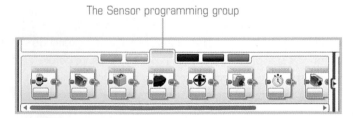

FIGURE 7.13 Grab your blocks from the yellow sensor group.

2. Find the Color Sensor block and drag it onto the programming palette. The Color Sensor block has a picture of the Color Sensor on it, but if you're ever confused about which block to select, just hover your mouse over the block, and the name will appear as a tool tip.

3. Drag the Color Sensor block onto the programming canvas just next to the Start block, as shown in Figure 7.14.

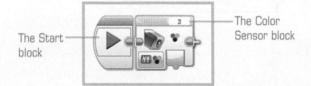

The Start block

The Color Sensor block

FIGURE 7.14 Drag the Color Sensor block onto the programming canvas.

In most cases, you'll want to put blocks right next to each other, because they're part of a sequence. They'll naturally snap together. Sometimes you may prefer to view your sequence all in one window and want to separate longer block chains to make it more convenient for your workflow, or you might be working on a group of blocks in another area. However, you can separate these blocks and still get them to work by using sequencing wires.

At the end of each block is a little gray tab with a bit of a stinger on the end, resembling perhaps a soldering weld. If you click and drag from that tab, your cursor will turn into an icon that looks a bit like a spool of wire. This is a sequence wire. You can then drag the sequence wire to connect one block to another block, as shown in Figure 7.16.

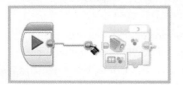

FIGURE 7.16 Connect blocks in a sequence by using a sequence wire.

For this program, keeping the blocks next to each other is fine.

Changing Modes

You now need to adjust the sensor block programming. The sensor block can actually do a huge number of things with your robot's sensor. By default, it is set to measure color. Detecting a black line might sound like something you need to do by measuring color, but it's actually much simpler to do this by detecting reflected light intensity. Follow these steps to choose the mode to do that:

1. Click the little measuring tool on the lower-left corner of the sensor block. This area is the Mode Selector (see Figure 7.17).

FIGURE 7.17   Click on the Mode Selector.

A drop-down menu offering the mode choices of Measure, Compare, and Calibrate appears. You could use Measure to tell you what color an object is or how light or dark it is, and it might work, but it is not the most efficient method to make this program work.

2. You want your program to compare the reflected light against a threshold to see if it is a black line, so choose Compare, as shown in Figure 7.18.

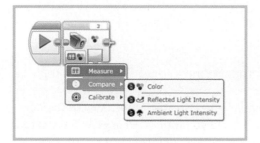

FIGURE 7.18   Choose compare.

Three more choices appear for you to select what factor to compare:

- **Color** compares colors. You could test to see whether something was red, for example.
- **Reflected Light Intensity** uses the sensor to shine a red light and check to see how much light is reflected back. This is an excellent way to check to see whether something near the sensor is black or white.
- **Ambient Light Intensity** checks to see how bright or dark the light is. Using this sensor is a better option if you just want to see how light or dark a room is. For example, you could use it to detect whether you would need to use a flash on a camera.

3. Click the Reflected Light Intensity option, because it makes the most sense for what we want the bot to do. Your block suddenly offers a lot more options, as shown in Figure 7.19.

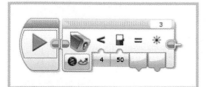

FIGURE 7.19   There are multiple ways to compare light.

The < symbol above the 4 indicates the type of comparison you want the sensor to make. If you click the 4, several other options appear that you could choose instead, as shown in Figure 7.20.

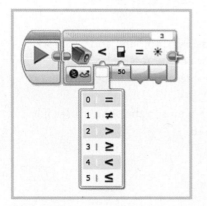

FIGURE 7.20 Your expanded choices for comparison.

As you can see, you can make all sorts of comparisons: Greater Than, Greater Than or Equal To, Less Than, and Less Than or Equal To. Do you want a value exactly equal to a threshold shade, such as all white or all black, for example? Do you want a value that is not equal to that threshold? The Not Equal option could be useful if you want the robot to go forward when the value is not black, for example.

In this case, you could opt to detect whether the reflected light is below 50, which is the default middle value. Because you want the robot to detect a black line and not just a darker patch of floor, set this value to 20. In Chapter 8, "More MINDSTORMS Programming: The Line-Following Robot," I discuss calibrating your sensor so that the threshold value is more accurate.

Checking Your Ports

On the upper-right corner of your sensor block is the sensor port number. Make sure it corresponds with where you've physically plugged in your sensor on your EV3. In this case, it should say 3, but if it doesn't, you can click on it and change it, as illustrated in Figure 7.21.

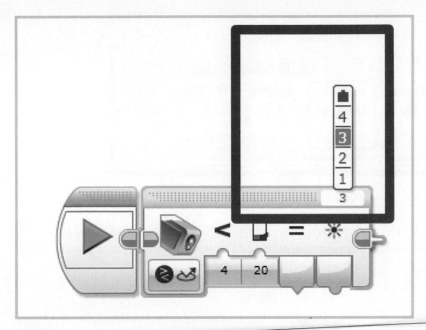

FIGURE 7.21 The sensor ports can be changed here.

Making the Bot Move

Now you need to figure out how to make your bot move. This corresponds with "Move the robot forward" on the flowchart from Figure 7.13. If you built your bot to the original design in Chapter 6, large motors are connected to ports B and C. (Remember that the alphabetical ports control motors and the numeric ports control sensors.)

The green tab on the programming palette contains action blocks; so let's start there. Multiple blocks could potentially control the wheels:

- The first is the Medium Motor block, and you can rule it out immediately because you're using large motors.
- The second is the Large Motor block, which sounds promising, but if you drag it onto the canvas, you'll see that it only controls one port at a time. That means you would need to have two Large Motor blocks that you must program at the same time.
- The third is the Move Steering block, shown in Figure 7.22. It is exactly the right block for your project.

FIGURE 7.22   Identify the Move Steering block by the little steering wheel.

The Move Steering block controls large motors on two blocks at once, and it makes the motors move or turn with the same power. Drag and connect it to the rest of your program in the canvas area.

> **NOTE**
>
> One more block you could use that's worth mentioning is the Move Tank block. It allows you to program two large motors at once, but it also gives you individual control over the power and motion of each wheel. It's like having two Large Motor blocks smooshed together. It's overkill for this project, because you just want your bot to move forward until it finds the black line.

Moving the Wheels

Take a closer look at the options on the Move Steering block (see Figure 7.23). The first option on the lower left lets you choose how you want to move the wheels. You can turn them off, or turn them on to keep going. You can time them for seconds, calculate how many degrees to rotate, or set a certain number of rotations. I like to use a single rotation, but a single second of power would work, too. You just want a small forward motion before the sensor checks to see if the robot has reached the line.

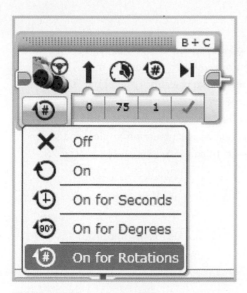

FIGURE 7.23 Set how many rotations, degrees, or seconds you want the wheel to rotate.

Controlling Bot Direction

The next option, shown in Figure 7.24, controls your bot's direction when the wheels move. Do you want it to go straight or turn to the left or right? The slider goes from -100 to 100 indicating how much power you want to give the right wheel as a percentage. If you want the robot to make a complete turn to the right, you'd set the value at 100, and a complete turn to the left would be -100. In both of those examples, there would be no motor power going to the opposite wheel. In this project, you want it to go straight, so set the value to 0. In future projects, you can use the slider to change a bot's direction.

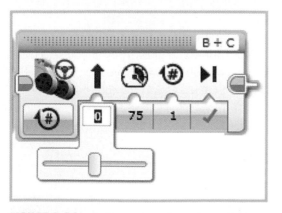

FIGURE 7.24 Adjust the direction of the steering to 0.

Adjusting Motor Power and Speed

After you've adjusted direction, you can adjust the power/speed of the motor (see Figure 7.25). Notice that the slider adjusts to a negative number, which makes the motor spin in the opposite direction. Generally, negative numbers make the robot travel backwards, and positive numbers make it travel forwards. In some cases, depending on how you've assembled the robot, you might consider "backward" to actually be forward. In this case, the robot should travel forward with a positive number. Just as with steering angles, the slider goes from -100 to 100, with 100 indicating full power. In this case, set the power to 75.

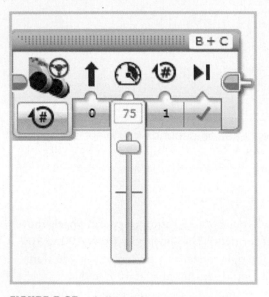

FIGURE 7.25 Adjust the motor power.

Adjusting Rotation Count

Next to the motor speed/power adjustment is the rotation count. This option changes as you alter how you opt to control your motors. If for your motor speed/power you select On for *Duration*, a stopwatch icon appears that lets you control how many seconds the wheel spins. If you choose On for Degrees, a circle appears and allows you to input how many degrees you want to spin the wheel.

The final option is what to do after the wheels have spun their specified rotation (see Figure 7.26). Since you rotated the wheel for one rotation, this is the option that appears. Your options after the specified rotation duration are to keep coasting or to stop.

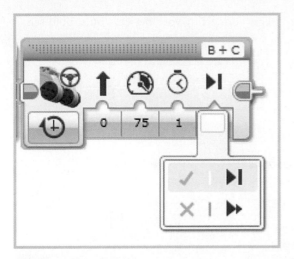

FIGURE 7.26   Tell the robot what to do once the wheels have spun.

For this project, select Stop, which is the first option and probably already selected by default.

Connecting Your EV3 to Your Computer

At this point you have a chain of three actions tied together, so you can try running the program. If you've already built and run some of the demo bots, you might already know how to do this. Feel free to skim.

NOTE

Even though there is a Start block, it isn't necessary to add a block to stop all the programs. EV3 programs end when they reach the end of the programming sequence.

The lower-right corner of your programming canvas shows you the connection status of your EV3 (refer to Figure 7.11). You can connect by USB cord from your computer to your EV3, Wi-Fi, or Bluetooth. If you have a computer that supports Bluetooth, then that is usually the easiest option. It's wireless, and it doesn't require you to have an Internet connection, so you could install and run a program on the road or at a tournament.

Here are the basic steps to run your program through Bluetooth:

1. Make sure your Bluetooth is turned on in your computer, and open the Bluetooth preferences.

2. Turn on your EV3, and press the right tab until you get to the wrench.

3. Scroll down to Bluetooth. Press down to begin connecting your device. You should see your EV3 as an option on your desktop, as shown in Figure 7.27.

FIGURE 7.27 Detect your Bluetooth devices.

4. Click the Pair button on your computer, and press the middle button on your EV3 at the same time. A code appears on your EV3. By default it is 1234. You only need to change it if you're somewhere where people have more than one EV3 they're trying to pair.

5. Enter the code on your computer when prompted, as shown in Figure 7.28.

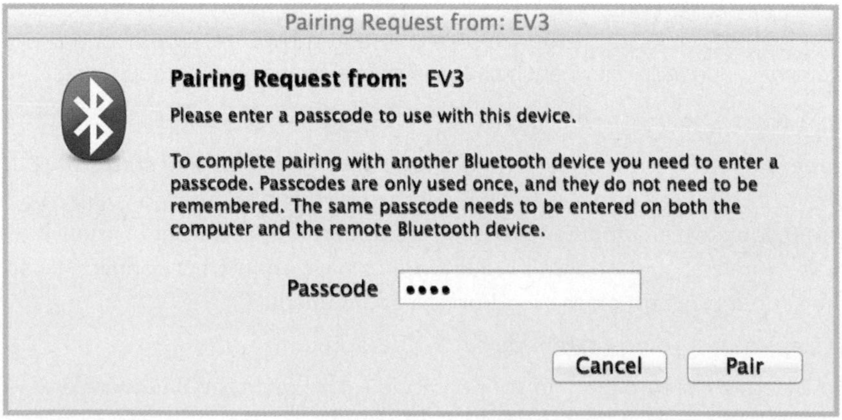

FIGURE 7.28 Here is a basic pairing request.

After you've paired your EV3 to your computer, you can connect the EV3 home edition software to your EV3 robot by clicking in the square under the Bluetooth symbol. Once your device is connected, you can run programs and debug your programs directly from the desktop. Figure 7.29 shows the lower-right corner of the EV3 software when it is connected to an EV3.

FIGURE 7.29   The EV3 is connected to the robot.

Go ahead and connect your EV3 to your desktop software now. This is a great way to test and debug your program as you go. When you run programs directly from your EV3 home edition software, the program highlights the area of the programming block that is active as the program progresses, so if there is a problem, you can identify where it occurs in the programming sequence.

To review, you should have a programming sequence that looks like Figure 7.30.

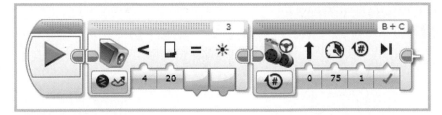

FIGURE 7.30   Your current programming sequence.

After you've connected your EV3 to your computer, click the play button on the lower-right corner of the EV3 home edition software, shown in Figure 7.31.

FIGURE 7.31   Click on the play button to run your program.

Your program should immediately start playing on your EV3, which means that most likely, the robot moved forward one rotation and then stopped. Try putting your robot on a black surface. Chances are it will do exactly the same thing. Why? Although you have told the EV3 to detect whether or not the reflected light was less than 20%, you didn't tell it what to do with the information. So instead of making a decision to only move forward under certain conditions, the EV3 just moves to the next command in sequence and moves forward one rotation. The program then reaches the end of the programming sequence and ends, so the robot also stops moving.

Making Decisions and Using Loops

You have a bot and a basic program. This program has two problems, however, relative to the initial flowchart. Let's take care of them.

The first problem is that you need to have the EV3 make a decision. One way to do that is to swap out the Light Sensor programming block for a Switch block:

1. Click and drag the Switch block from the orange palette tray, as shown in Figure 7.32, and place it onto the canvas.

FIGURE 7.32   Use a Switch block.

2. Remove the Light Sensor block, by clicking on it and pressing the Delete key. You probably noticed that when you dragged the Switch block onto the canvas, it expanded a lot and gave you two choices, one with a check and one with an X, as shown in Figure 7.33.

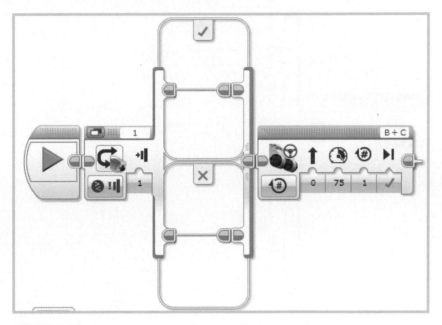

FIGURE 7.33   *The expanded Switch block is shown.*

The check is for when the condition is true, and the X is for when the condition is false. You'll also notice that the picture on this switch is of the Touch sensor.

3. Click the button just below that Touch sensor icon and switch it to a Reflected Light sensor just as you did for the Sensor block in the section, "Changing Modes." You should also change the threshold to 20 and the mode to 4. See Figure 7.34.

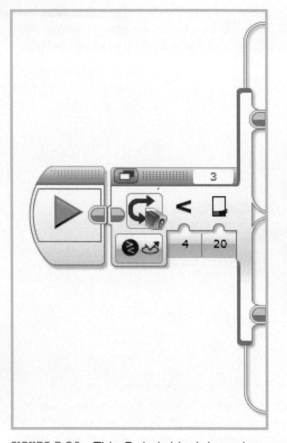

FIGURE 7.34   This Switch block is acting on sensor information.

Remember, this setting tells the sensor to detect whether or not there is less than 20% light being reflected back. In other words, is the sensor over something dark? If the answer is yes, the bot should stop. If the answer is no, it should keep going.

4. Drag your Move Steering block (the green block in Figure 7.34) to the X or the NO position because you only want your bot to move if the answer is no. Because you don't need it to do anything when the answer is yes, you don't need to drag any blocks to the YES position (see Figure 7.35).

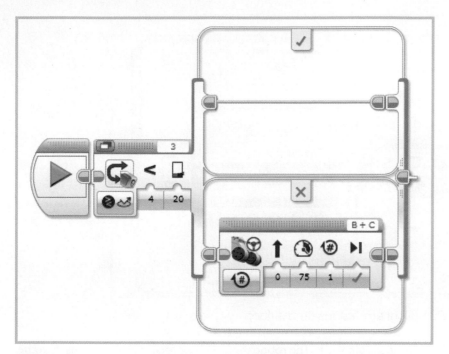

FIGURE 7.35 In this case the "yes" condition should do nothing.

Try your program now. When you press the play button to run this program on your EV3, the bot should correctly go forward when it is on a light surface and not go when it is on a dark surface. This brings us to the second problem relative to the flowchart: The bot won't keep going. Even changing the setting to not brake at the end won't fix this. What you need at this point is a loop. A *loop* keeps the programing blocks in it going for as long as the loop is active. By default, the loop goes on forever. To set a loop:

1. Back in the orange Flow Control section of the palette, drag a Loop block onto the canvas.

2. Drag your switch into the loop. It might look too small, but the loop can expand to hold whatever you drag into it. It should look something like Figure 7.36.

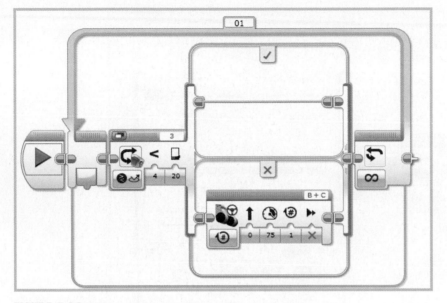

FIGURE 7.36 Our program fits inside the loop.

Try your program now. It should work. The robot should move forward until it runs into the black line, and then it should stop moving. The program will keep running, but the bot will stop moving. In fact, you'll have to click the stop button on your computer to stop running the program.

TIP

As you try out your loop, pay attention to your computer screen. Each block sparkles a little to indicate when the program is executing the commands on that block. This feature can help you debug your program or make it more efficient.

At this point you might think your work is done, but you can make this program even more efficient. What if you used the Loop block itself as the way to sense the line? To do that, follow these steps:

1. Drag your green Move Steering block out of the Switch block so that it connects next to the Loop block.

2. Delete the Switch block.

3. Notice how the Loop block also has a Mode Selector button. By default it is set to Infinity, but if you click on it, you can see all sorts of choices, including the reflected light sensor.

4. Adjust the mode to 4 and a threshold of 20; your program should look something like Figure 7.37.

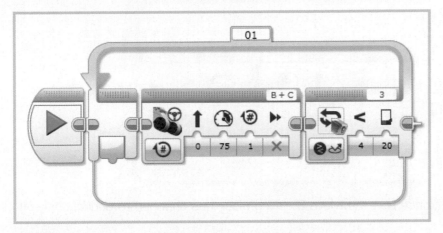

FIGURE 7.37   Our simplified program uses a loop.

Now do you see what is happening? The loop will keep going, as long as the condition is NOT true. The condition is that the light sensor is detecting reflected light of less than 20. As soon as it detects a dark patch, the loop exits, and then nothing else happens because the program has run out of programming blocks to execute.

Go ahead and play this program to test it. It should now work exactly like the flowchart shown earlier in Figure 7.13, and it uses only three blocks. You might notice one final problem with the bot. It tends to stop and start. There's an easy fix for that. Because the program no longer contains a loop, you can just change the Move Steering mode to continuously on (see Figure 7.38).

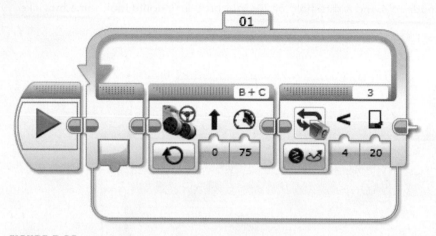

FIGURE 7.38   Notice how the Move Steering block has changed to continuously on.

Saving Your Changes

Now that you've made your first program, be sure to save your changes. You can either choose File, Save Project, as shown in Figure 7.39, or click the disk icon.

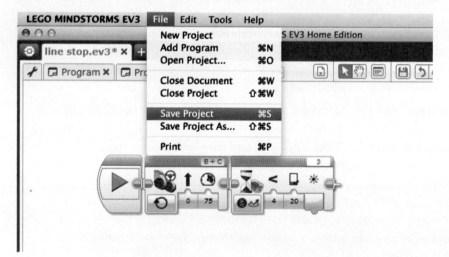

FIGURE 7.39   Save your changes.

You'll need to give your new project a name. I chose line stop.ev3, but you can choose whatever will help you remember. By default, it's saved in your Documents folder. You can share this .ev3 file with others if you want. You can also document (or add Comments to) your program if you want to leave yourself or others notes about how your program works.

Alternative Programs

The example program you've just built isn't the only way you can program an EV3 to stop on a line. In fact, there are probably dozens of ways to make the program work. Figure 7.40 shows another solution, using the Timer block instead of a loop.

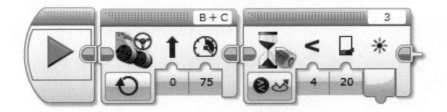

FIGURE 7.40 Here's the same program with a Timer block.

In this case, there is a timer running after the Move Steering block has turned on the motors. The timer tells the program to keep going until the light sensor sees the dark line and then do the next thing. The next thing is nothing. This is important to note. Your EV3 can execute multiple instructions at the same time. It goes on to the next programming block after it turns the motor on and doesn't wait for the first command to finish.

Try running this program *without* the Timer block. What happens? Your program will most likely stop before it starts moving forward.

Summary

In this chapter, you got started with EV3 programming. You created a program that used the light sensor and the large motors of your EV3 to stop on a line. In the next chapter, you learn more advanced programming that makes the sensor actually steer and follow a line.

More MINDSTORMS Programming: The Line-Following Robot

In Chapter 7, "Make Your First EV3 Program," you programmed a modified Educator bot to stop on a black line using the color sensor. In this chapter, you'll continue using the same basic robot with the color sensor to delve a little deeper into more things you can do with your EV3, such as change the screen display on the EV3 and make sounds using the built-in EV3 speakers. To that end, this chapter covers some more advanced programming concepts that you can use with your EV3, such as variables.

TIP

The information you learn in this chapter can also be generalized beyond LEGO and used for many different kinds of programming.

In Chapter 7 you made a program, but what does that mean? Let's go a little deeper.

What Is a Program?

Have you ever played the classroom game where you tried to tell an alien how to make a peanut butter and jelly sandwich? This is an alien who is visiting Earth for the very first time. Assume in this case that your alien knows perfect English but learned it all from a dictionary and does not understand figures of speech such as, "I'm so hungry I could eat a horse." This alien will take everything you say literally, so be careful with your commands.

So, to teach an alien to make a classic PBJ, perhaps you would start with something like this:

1. Get two slices of bread, a jar of peanut butter, a knife, and a jar of jelly.

2. Put peanut butter on one slice of bread.

3. Put jelly on the other slice of bread.

4. Put the two slices of bread together.

Simple enough, right? Well, your alien interprets everything literally, so your sandwich will probably look something like Figure 8.1

FIGURE 8.1   Here is the alien's interpretation of peanut butter and jelly on two slices of bread.

You see what is going on with this picture? The literal alien put the peanut butter on one slice of bread, the jelly on the other, and then put the slices together. You should have been a little more specific, maybe with something more like this:

1. Get two slices of bread, a jar of peanut butter, a knife, and a jar of jelly.
2. Open the peanut butter jar.
3. Use a knife to scoop out one tablespoon of peanut butter.
4. Put peanut butter on the face of one slice of bread by spreading it with the knife.
5. Clean the knife.
6. Close the peanut butter jar.
7. Open the jelly jar.
8. Use the knife to scoop out one tablespoon of jelly.

9. Spread the jelly on the face of the other slice of bread using the knife.

10. Clean the knife

11. Close the jelly jar.

12. Put the peanut butter and the jelly faces of the bread together.

Such detailed steps require a lot more effort to write, but they are also much clearer to the alien who must follow them. Even then, some confusion might still arise, and you might need to rewrite your instructions a few times to make sure the alien gets it exactly right.

Writing a computer program is a similar process. You're giving instructions to a robot that will follow every one of your commands literally. It has pre-defined vocabulary that you can use to give these commands, and you will spend a lot of time troubleshooting things that you think should have been obvious and easy. If there's one thing in this book I cannot repeat often enough it is that programmers need to be resilient. Keep testing, keep trying, and keep making new iterations of your design.

> **NOTE**
>
> When you program, you'll usually end up making a lot of versions with incremental improvements as you go. Don't ever expect your prototype to work perfectly the first time.

Project: The Line-Following Robot

In Chapter 7, you programmed a robot that stopped on a line. This project takes this a step further and makes the robot follow a black line around a test track. It needs to steer and correct itself as well as detect the difference between the black line and white track background. For this project you use the same basic modified Educator robot car you constructed in Chapter 6, "Hacking What You Have" (see Figure 8.2).

FIGURE 8.2   This is the home edition version of the basic robot, but you can also use the LEGO Education version.

As a refresher, this robot has two large servos powering the wheels and a color sensor attached in front. Your project is to make the robot follow a line. It must stay on the line and not get lost by going so far off course that it cannot find the line. This is a classic robotics exercise and a common challenge for robotics contests.

Getting Started

You need to do a little prep work to get started on this project. The EV3 home edition comes with a test track (see Figure 8.3), but that track doesn't have a black line on it, and it doesn't have a circular path, so the robot would leave the test track pretty quickly.

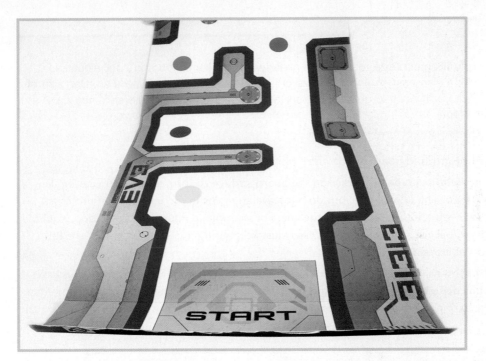

FIGURE 8.3 You could use the test track from the EV3 home edition box.

You can approach this in one of two ways: Make the EV3 follow a red line on the test track, or make your own test track with a black line. Because making your own track means you can make a loop and watch your robot follow the line for much longer, let's try that approach.

Making Your Own Test Track

To make your own test track, follow these steps:

1. Get a large trifold presentation board. It's a cardboard display that you can use for things like science fair presentations. When you unfold it, the inside is white. You can buy them at office supply stores or retail outlets like Target. I purchased one for $2.50.

2. Find either black electrical tape or a very thick marker. The ¾-inch width electrical tape will work well for this. Black electrical tape is sticky, but not so sticky that you can't pull it up and make adjustments on most surfaces.

> **NOTE**
>
> The Sharpie Magnum size marker is a great wide marker if you can't get electrical tape, and it's also easy to find in stores or online. You could also use a smaller size of marker, but you'll want to go over the surface several times to make sure you have a thick, dark line.

3. Unfold your trifold board and put the white side face up.

4. Use the electrical tape or marker on the board surface to form a loop. You can make waves or corners in your line, but do not leave any gaps in the line, and do not cross the path with another segment of the line. For example, a figure 8-shaped loop would not work, because the robot would likely just keep going in a circle in half of the loop instead of crossing it.

Figure 8.4 shows the homemade track I made with a wide-tip marker. This track illustrates two options depending on your comfort level with line tracking. The more challenging track, featuring sharp corners and tricky bends, is on the outside. The easy track, an oval, is on the inside.

FIGURE 8.4 Here are two lines—one easy, one challenging—you could practice following with your bot.

Thinking About the Instructions

Let's return to the peanut butter and jelly analogy from the beginning of this chapter. What is the list of instructions you need to give your robot to make it complete the mission of following the line? Maybe something like these:

1. Go forward when you're on the line.

2. Turn toward the line if you start to veer away from it.

3. Repeat these instructions.

These instructions are pretty good, but they need to be translated into something a little closer to how the EV3 follows instructions. Let's try this:

1. Use the color sensor to detect reflected light.

2. While the sensor detects the line, power both large motor servos forward at the same rate.

3. When the sensor does not detect the line, turn one large motor servo on the robot until the line is detected.

4. Repeat these instructions (see Figure 8.5).

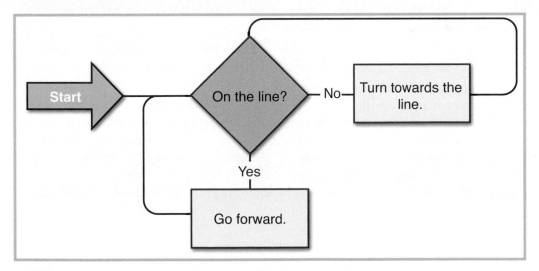

FIGURE 8.5   Here is an example flow chart for these instructions.

Finding Direction

A challenge for keeping a robot following a line is this: How can you tell which way to turn the robot? It could veer to the left or to the right of the line. Do you spin the robot in one direction until it hits the line, even if that ends up spinning it in a nearly complete circle? That's one possibility. Do you turn it a little to the left and then a little more to the right and then a little more to the left and so on until it finally finds the line? That's another possibility.

Some people also solve the problem of knowing which direction the bot should turn by using three separate color sensors right next to each other. That way one sensor is on the line, one is to the left of it, and one is to the right of it. Your EV3 kit only has one sensor, so let's work on a solution that doesn't require more.

Here's a suggestion that will make things a lot easier: Straddle the line. Instead of trying to put the color sensor in the exact center of the black line, put it on the edge.

Suppose you put the color sensor on the edge so that the left half of it is on the black line and the right half is on the white area in between. Now, instead of detecting a nearly black line, the proper reading should be somewhere around a 50% gray. When the sensor detects more than 50% darkness, you know that the EV3 has strayed too far to the left. When it detects more than 50% light, you know it has steered too far to the right. You can correct course immediately with micro adjustments instead of making the robot spin all over the place to find the line. The motion would look something like what's shown in Figure 8.6.

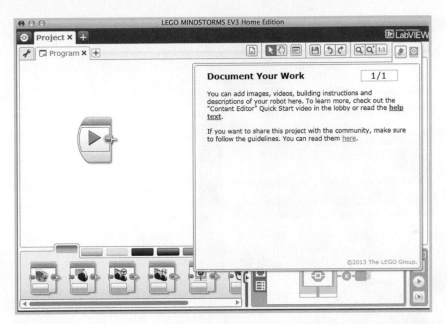

FIGURE 8.6   This is the classic zigzag motion of many line-following robots.

Calibrating the Sensor

Straddling the line at 50% produces one more problem. Your homemade track might not be perfectly black or perfectly white, and you might not have the same lighting conditions every time you try out your robot. So what value should you set for your sensor to make sure it is 50% of your lighting conditions? You need to calibrate your sensor to match your environment. Here's how you manually do this:

1. Power on your EV3.

2. Press the right navigation button twice (see Figure 8.7) to navigate to the third tab.

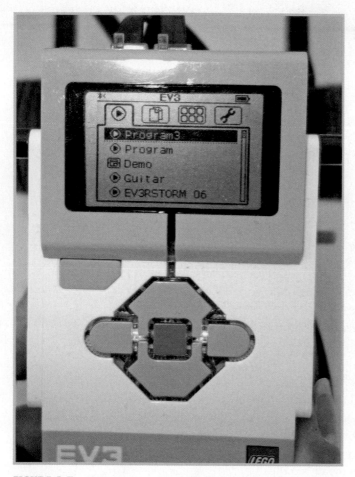

FIGURE 8.7   Press the right navigation button twice.

3. Select Port View and press the center navigation button (see Figure 8.8).

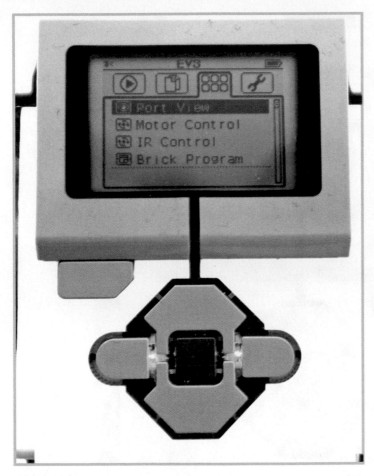

FIGURE 8.8 Press the center navigation button to select Port View.

4. Navigate to the port that has your color sensor. In this case, it is port three, so press the right navigation button twice (see Figure 8.9).

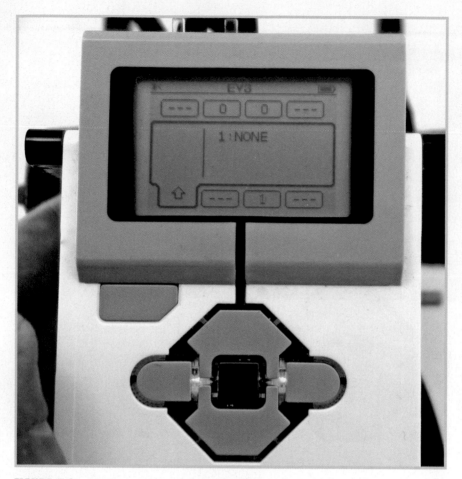

FIGURE 8.9 Navigate to the third port.

You'll now see the color sensor. It is in Reflect mode by default (see Figure 8.10).

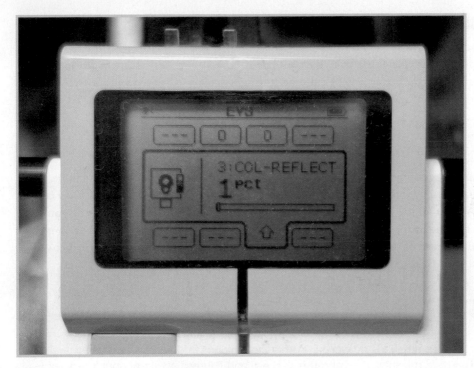

FIGURE 8.10 Now you can calibrate your sensor.

5. Place your robot on the track. Place the sensor over the blackest area of your line, and then put it over the whitest area. You'll see the reflect value (as a percentage) change over each area. Write down the sensor value for both of these areas.

6. Add both those sensor numbers together and then divide the result in half, or $(a+b)/2=n$. That number (n) is your midpoint gray. That's the number your color sensor should register whenever it is perfectly aligned on the edge of that black line.

> **NOTE**
>
> Remember when programming that you indicate division with a slash and multiplication with an asterisks, so one divided by two is written as $1/2$ and one multiplied by two is written as $1*2$.

Keep in mind that with manual calibration, you have to do it for every new environment in which you use your EV3, because the lighting might be different. The better solution is to have the sensor calibration be part of the program itself, which means moving over to the EV3 software on your computer.

Creating the Program

You have your test track with its white background and black line. Now it's time to start building the program you'll use to make your robot do its thing. Let's start by learning a more intricate, but far better way to calibrate the light sensor—so you don't have to spend time checking values as you move your robot from one environment to the next. To do that, you need to create a variable.

Creating New Variables

Variables are programming elements that can change. They're like buckets that can be filled with something different each time you run the program. Think of the lyrics to Happy Birthday. Almost all the words to it are always the same, but you change the name you sing to be specific to the person who is celebrating the birthday. The name in that song is a variable. In the case of this project, you want the variable to be the midpoint gray between the lightest and darkest areas.

> **NOTE**
>
> Variables can be numbers or words. In this case, your variable is a number representing the midpoint between the darkest and lightest areas of your track, or $(a+b)/2 = n$.

So let's get started programming the calibration tool. You might want to save this program by itself, instead of as part of your robot's full programming, so you have a handy calibrator available for other programs and builds.

1. Open the EV3 home edition software on your desktop computer.

2. Click File, New Project to launch a new project.

> **NOTE**
>
> If you haven't yet worked with the EV3 software, or if you find yourself getting lost following these instructions, make sure you've thoroughly reviewed Chapter 7.

3. The Content Editor is open by default, but it will get in your way as you drag items onto the canvas, so click its close button.

4. Drag a Wait block onto the canvas from the orange block group (see Figure 8.11).

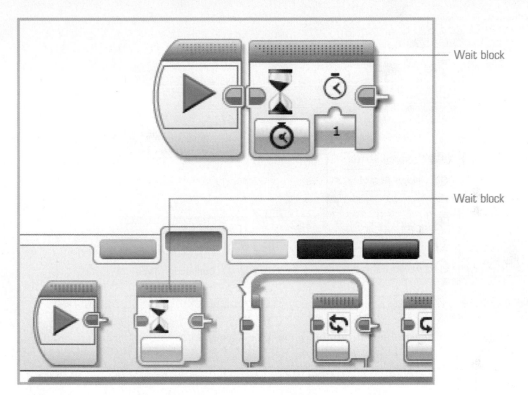

Wait block

Wait block

FIGURE 8.11 Drag the Wait block, and more options appear.

5. The Wait block can wait for all sorts of events. In this case, you want to wait until you have your sensor positioned over the white area of your test track. The simplest thing to do is use the buttons on the EV3 itself. There's no reason to be specific. Let's have the Wait block wait until any button is pressed. To do that, set the bottom-left section of the Wait block to Brick Buttons, Change, Brick Buttons as shown in Figure 8.12.

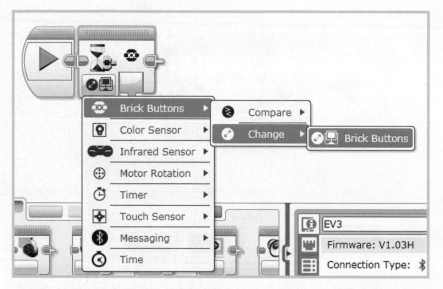

FIGURE 8.12   This setting waits for any Intelligent Brick button to be pushed.

6. Drag a Color Sensor block from the yellow block group onto the sequence (see Figure 8.13).

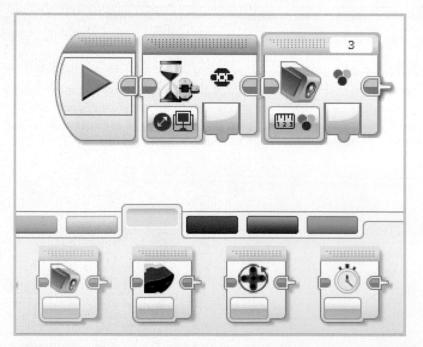

FIGURE 8.13   You need to adjust the Color Sensor block to measure reflected light intensity instead of color.

7. Set the Mode Selector to Measure, Reflected Light Intensity (see Figure 8.14).

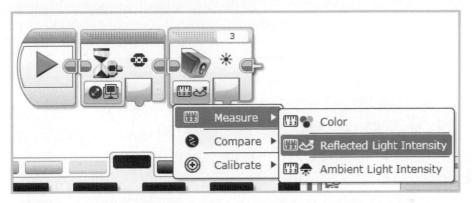

FIGURE 8.14 Set the Mode Selector to Reflected Light Intensity.

> **NOTE**
>
> If the goal is to calibrate the sensor, you might wonder why you don't choose
> Calibrate. The Calibrate setting tells the sensor what the lightest or darkest point
> *should* be. So if you know your darkest point is 80%, you can calibrate the sensor to
> think 80% is actually 100%. Right now you don't know what the lightest and darkest
> points are, so you must measure those points first.

8. Drag a Variable block onto the sequence, as shown in Figure 8.15. The variable block
has a suitcase or briefcase icon, because it is a container that can contain what you
want it to contain.

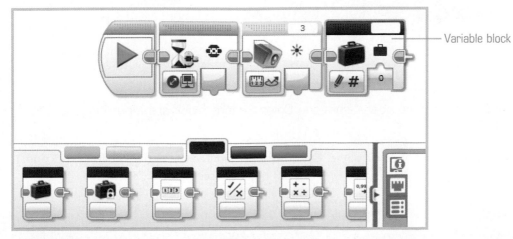

Variable block

FIGURE 8.15 This is the Variable block.

9. You need to give the Variable block a value. To do that, you need to connect the output value of the Color Sensor block to the input on the Variable block. You do that by using a data wire. This is similar to a sequence wire, only instead of connecting blocks to each other in the sequence, it connects data inputs and outputs to each other. Hover your cursor over the output value on the Color sensor, as shown in Figure 8.16.

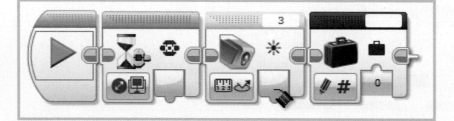

FIGURE 8.16 Notice how the cursor changes to a data wire symbol when you hover over the output value area.

10. Drag the data wire between the output value of the sensor and input value of your variable (see Figure 8.17)

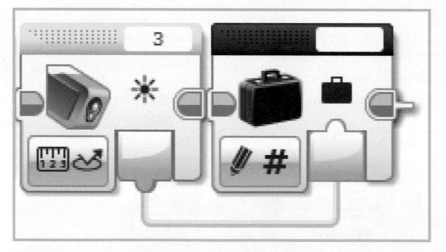

FIGURE 8.17 The numbers from the Color Sensor block are input into the Variable block.

11. Give your new variable a name. Click the upper-right corner of the Variable block and enter the name **Black** in the window that pops up (see Figure 8.18).

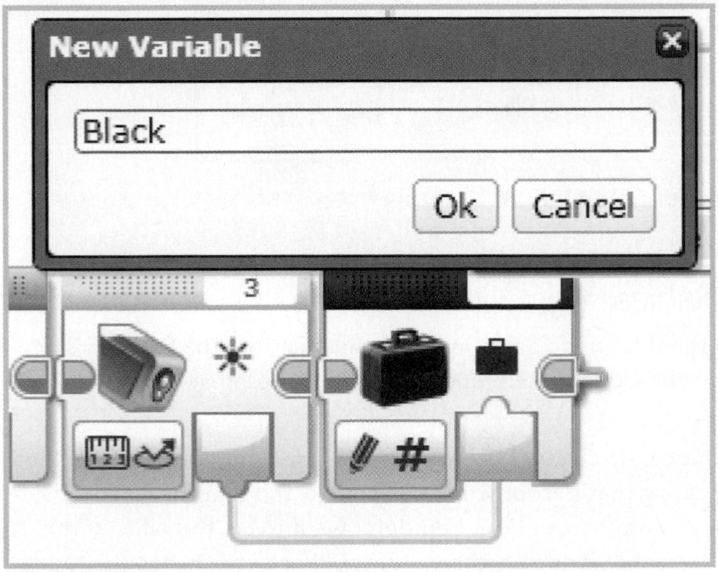

FIGURE 8.18 This is how you name your variables.

12. The steps for getting the white value are exactly the same as for getting the black value, other than the name of the variable. Rather than drag and tweak all those blocks onto the canvas again, you can simply copy and paste what's already there. Hold down the Shift key and click on the color sensor and variable blocks individually to select them, and then press Ctrl+C (Windows) or Command-C (Mac) to copy those items.

13. Press Ctrl/Command+V to paste a duplicate of the three blocks onto the canvas.

> **NOTE**
>
> You can also use the Edit menu to select Copy and Paste if you don't want to use the keyboard shortcuts.

14. Drag the three duplicated blocks onto the end of your programming sequence.

15. Change the name of the copied variable to White by clicking on the upper-right corner of the copied variable's block (which says "Black") and choosing Add Variable.

Your sequence should look similar to Figure 8.19.

FIGURE 8.19   A basic calibration sequence.

Calculating with Variables

It's almost time to use those variables to do some math. Remember that the formula is (a+b)/2. Drag another Variable block onto the sequence. It should automatically be set to Black.

You also have to change the mode. By default it shows a pencil and a hashtag (refer to the variables in Figure 8.19), indicating that it is for writing a variable and that the variable will be a number. That isn't what you want to do here. You don't want to set the value of the variable, but rather read the value that is already there. So, you'll need to change the Mode Selector to Read, Numeric (see Figure 8.20). The icon should change to an open book and a hashtag, indicating that it is reading a number variable.

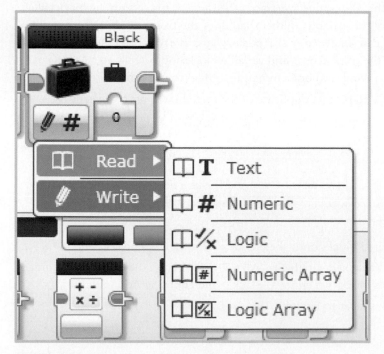

FIGURE 8.20   Set the node to Read-Numeric.

Repeat these steps by adding another Variable block, changing the mode, and then changing the variable name to White.

With the variables in place, it's time to do that math:

1. Drag a Math block onto the sequence from the red programming block group (see Figure 8.21).

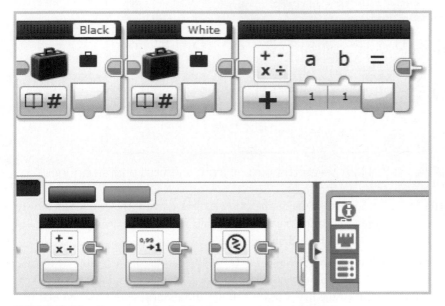

FIGURE 8.21 Drag a Math block onto the sequence.

2. Following the order of operations, you first add a+b in the (a+b)/2 formula. By default the Math block is set in addition mode, so you can simply use the data wire to connect the Black and White variables into the a and b inputs (see Figure 8.22).

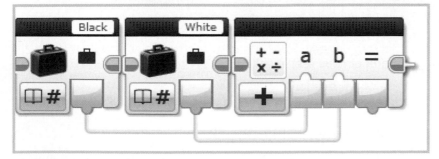

FIGURE 8.22 This Math block adds the number values of the Black and White variables.

3. Divide the result by dragging another Math block onto the sequence and changing the mode to Divide.

4. Drag a data wire from the = output on the first Math block to the input on the second one. The block should now look similar to the one shown in Figure 8.23.

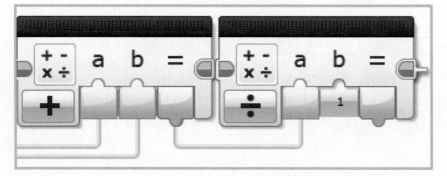

FIGURE 8.23 This Math block takes the output from the first one as an input.

5. Right now the Math block is set to divide the value it receives from the initial block by a value of 1. That isn't going to change the number at all. You need to divide it by 2. Click on the number 1 in the input of b, and change it to a 2.

6. The output of this Math block will be half of a+b, so you can use it to create the third variable. Drag another Variable block onto the sequence, connect the data wire from the second Math block output to the new variable input, and rename your block **MidGray**. You should get something that looks like Figure 8.24.

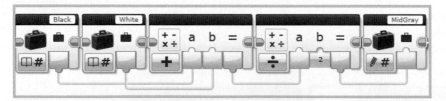

FIGURE 8.24 Your MidGray variable is now set to the midpoint between the white and black sections of your test track.

Ta da! You've made a basic calibration sequence. Now let's do a few things to improve it.

Improving the Program with Feedback

Although you have a complete and working program, it can still be improved. For example, one problem is that although this program does work, it's hard to tell it's working. There's no indication you should press the button, and you might forget whether you're supposed to scan the white or the black section first. (Technically the program would work if you reversed the order, because all the program does is add the two values and divide them in half, but it is still a good habit to give user feedback.)

Let's create some feedback with both some sounds and some screen changes.

1. Drag a Display block onto the canvas and insert it right after the Start block, as shown in Figure 8.25.

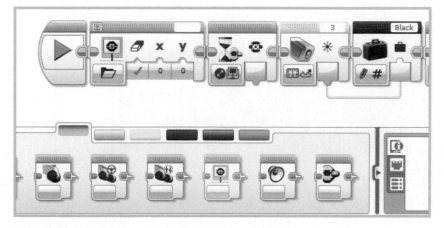

FIGURE 8.25 Drag a Display block onto the sequence.

2. By default, the Display block is set to use a file, but all you really need in this case is text. Change the mode to Text, Grid, as shown in Figure 8.26.

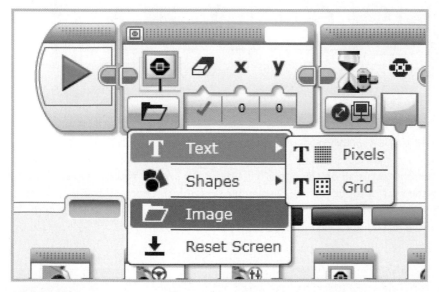

FIGURE 8.26 Set the mode to Grid.

3. The text in the upper-right corner indicates what will be shown on the screen. By default the text is set to MINDSTORMS. Click on the upper-right corner of the block and change it to **Black**).

4. Test to make sure your text is displaying by clicking on the little display icon on the upper-left corner (see Figure 8.27).

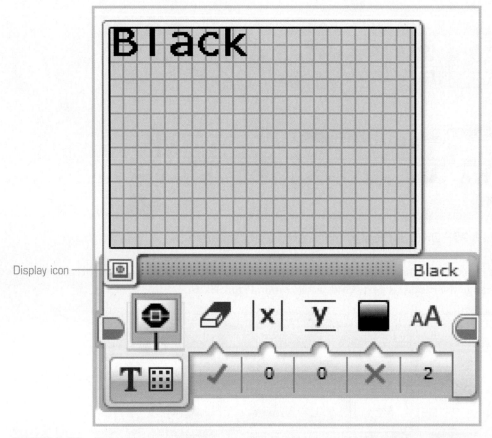

FIGURE 8.27 The display preview.

5. As you can see, the text is all on the upper-left side of the screen. You can tweak this text by using the x and y inputs on the Display block. You can change the size of the text using the AA input on the bottom right of the block. Tweak the numbers until you're satisfied with the results. One example is shown in Figure 8.28.

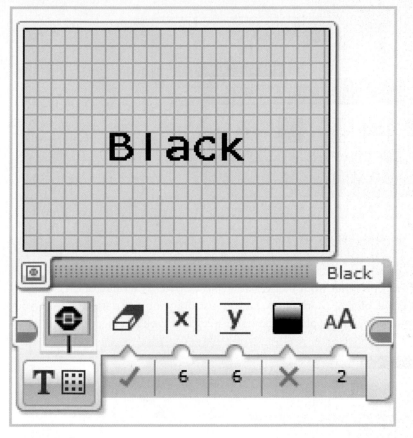

FIGURE 8.28 Here, the text is centered.

6. Drag another Display block (or copy and paste this one) just before your second Wait block, as shown in Figure 8.29. Be sure to rename it **White**.

FIGURE 8.29 Both the black and white screen prompts are set up.

Now that you have those prompts set up, wouldn't it be nice to know what the actual value of the midpoint gray is? It might help you troubleshoot in case something has gone wrong with the program. You can have a new Display block show the final MidGray variable:

1. Drag a Display block onto the end of the sequence. Change the mode to Text, Grid. Now, rather than changing the default MINDSTORMS value to some other text item, change it to Wired (see Figure 8.30).

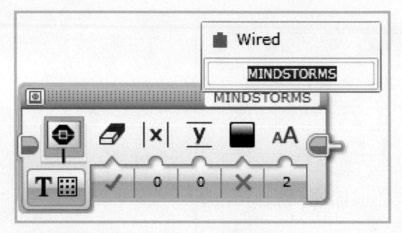

FIGURE 8.30   Change the display to Wired.

2. This immediately gives you another input. But wait, you just made the variable for MidGray, and there's no input from that variable. No worries. The value of the input came from the output of your Math block, so you just drag another data wire from your Math block output into your Display block input. It should look like Figure 8.31.

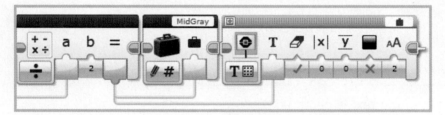

FIGURE 8.31   Connect the data wire to change the display.

3. Add a Wait block to the end of this sequence and set it for a 30-second delay. This is for testing purposes. As soon as the program ends, the display changes, so if you don't pause, you'll never see the display.

Troubleshooting Your Program

Go ahead and test your program. Did it work? If you followed the instructions exactly, you should see the brick display "Black" until you place the EV3 over the black line and press the button. It will then display a low number. It skips the white step completely. What happened?

What happened is that the program executes faster than you do. The Wait buttons wait for the buttons to change. Pushing down a button is a change. Releasing a button is also a change. When you press down on the button on the Intelligent Brick, the program has already gone ahead to the next Wait block, which means releasing the button triggers the color sensor to scan for the white portion of the track. How can you fix this? You can use one of several approaches:

- Hold down the button and carefully move the EV3 to the white portion of track. This doesn't seem very practical.
- Put another Wait block right next to the first Wait block on each sequence, so it actually takes two actions—pushing the button down and then releasing it.
- Change the Wait block mode to Brick Buttons, Compare and then specify an event on a specific button (see Figure 8.32)

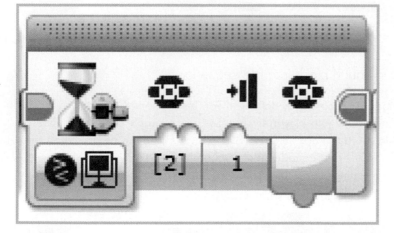

FIGURE 8.32   The Wait block is set to trigger only when the center button is pressed.

- Add another block to the programming sequence that would delay the program from reaching the next Wait block long enough to allow you to release the button.

Let's go with the last option. Let's also add a Sound block to the sequence. A sound prompts the user to investigate why the robot is beeping and make her more likely to look at the screen. It takes longer to make a sound than it does to release a button, so the sequence will work after you fix this. It's also a great excuse to explore the Sound block. Follow these steps:

1. Drag a Sound block onto the canvas and place it between where you defined your Black variable and where your screen starts to display the word *White* (see Figure 8.33).

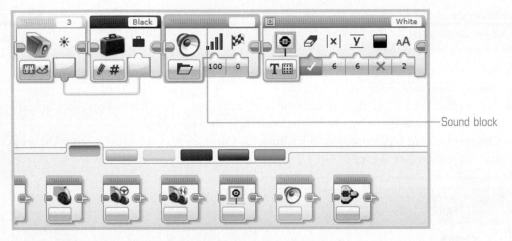

FIGURE 8.33 Place your Sound block into the sequence.

2. The Sound block's mode is set to File by default. If you have a sound file you've created or imported, this is how you would play it. In this case, use a simple tone to indicate that the robot has calculated the first variable (Black) and is ready to be placed on the white area of the track. Change the mode on the Sound block to Play Tone. You might also want to change the timing from 1 full second to .5 seconds (see Figure 8.34).

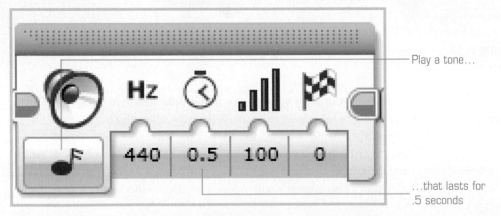

FIGURE 8.34 Adjust the Sound block to play a tone at 440 Hz for half a second at full volume.

Test your program again. This time, it should work. You should see a value at the end that is exactly between the lightest and darkest points on your track, generally a number between 30 and 60.

One last thing before moving on: I initially had you set up two Math blocks to show you how to make the calculation for your MidGray variable. I did this because it keeps things simple and it's the best way to learn what's happening in the program, but you can actually do this with just one block:

1. Rather than setting the mode of one block to Add and the other to Divide, take a single Math block and set the mode to Advanced.

2. On the block text field in the upper-right area, type in the formula **(a+b)/2**.

3. Connect the data wires to your variable and screen display (see Figure 8.35).

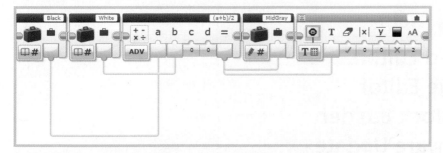

FIGURE 8.35 A single Math block now does the job of two.

When using the Advanced mode on Math blocks, you can ignore unneeded inputs (c and d in this case), but it gives you enough options to perform quite a few algebra equations.

Adding a Countdown

Now that the sensors work, let's give your line-following robot a countdown until it starts running. You can use the screen to display 3, 2, 1. If you do this, you also need to put a delay into the sequence with Wait blocks or some other action, or else you'll run into the same problem you did when you first tested your calibration tool—the countdown would happen so fast that you wouldn't see it at all. Let's use the Sound blocks again, only this time let's make a custom sound file.

1. Add a Wait block on the end of your sequence. Change the timing to 5 seconds. That should be long enough to see the value of MidGray.

2. Drag a Display block to the end of the sequence and change the mode to Text, Grid.

3. Change the text to 3, the x value to 9, and the y value to 6. It should look like Figure 8.36.

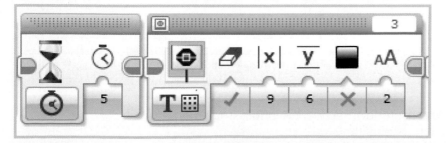

FIGURE 8.36 Set your display properties for this block.

4. Add a sound block and set the mode to Play File.

5. Click on the text entry area on the upper-right corner of the block. You should see a window that allows you to choose from Wired, Project files, and LEGO Sound files. Choose LEGO Sound files (see Figure 8.37).

FIGURE 8.37 Choose LEGO Sound Files.

6. Scroll down and select the folder called Numbers and then the file called Three. You now have the basics for your countdown.

7. Copy and paste the Sound and Display blocks two more times on your programming sequence. (You should have three consecutive sets of Sound and Display blocks.)

8. Change the numbers on the second and third Display blocks to 2 and 1, respectively.

9. Change the files on the second and third Sound blocks to match. The end of your sequence should look like Figure 8.38.

FIGURE 8.38 The countdown sequence is fully implemented.

TIP

Just for fun, you can add a sound block that says "Go!" You can find that file in the Sound Block's Communication folder.

Using a Loop to Make the Robot Follow the Line

Ok, now that you've calibrated the sensor and given your robot a countdown, it's time for the final portion of the sequence for this line-following robot. You need to make a continuous loop that checks the light intensity. When it is too light, the robot should turn left. When it is too dark, the robot should turn right. This makes the robot have that zigzag motion I described in Figure 8.6 much earlier in this chapter.

Let's get started:

1. Drag a Loop block onto the end of your sequence. This loop contains everything else in this last canvas of the build (see Figure 8.39).

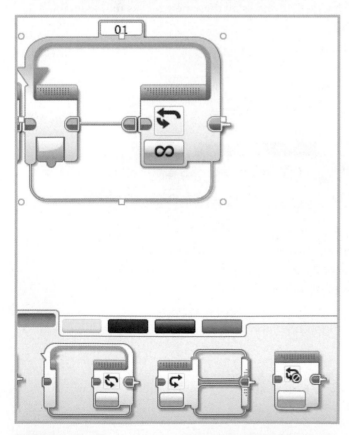

FIGURE 8.39   Drag a Loop block into the sequence.

TIP

You can name loops. This is particularly helpful when you have more than one loop in a program or you're trying to remember why you made a loop. Just click on the number on the top of the loop and give it a name. I named this loop "Line Seeking."

2. Drag a Variable block into the loop. Change the mode to Read, Numeric and select MidGray as the variable.

3. Add a Color Sensor block after the Variable block. Set the mode to Measure, Reflected Light Intensity.

4. Add a Compare block to the sequence. The Compare block can make comparisons between numbers. Set the mode to Greater Than.

5. Use data wires to connect some of these elements. You're trying to see whether the Color sensor is detecting a reading that is a higher number than the MidGray value, so drag the data wire to connect the Color Sensor block output to "a" and the MidGray Variable block output to "b." Figure 8.40 shows all these new blocks fully assembled.

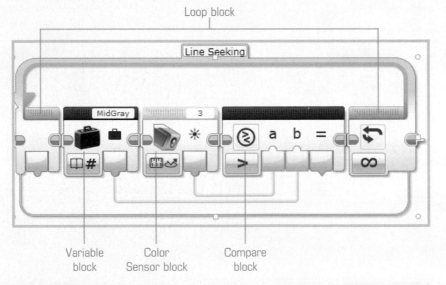

FIGURE 8.40 The Line Seeking loop now contains Variable, Color Sensor, and Compare blocks.

Adding Switches for Steering

The output on the Comparison block is going to be one of two things: True or False. Either the sensor has a higher number or it does not. In this case, the two numbers being exactly equal would still be False. So what you need to do now is have two choices of how to react. You do that with the Switch block:

1. Drag a Switch block onto the sequence within the loop. The loop should expand to accommodate it (see Figure 8.41).

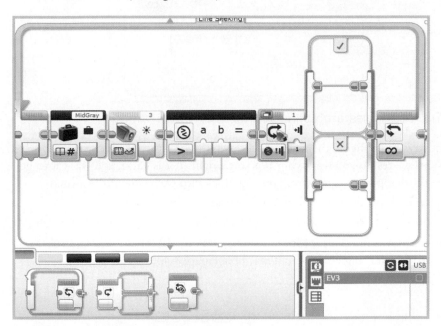

FIGURE 8.41   Insert the Switch block.

Switches are important programming elements, because they allow you to make choices. Notice the check on one side of the switch and an X on the other side. This is for True and False.

2. Before you go any further, you need to change the mode on the Switch block. By default it is set to Touch Sensor. This is so you can use the touch sensor as a literal switch. You're not using the touch sensor in this part of the program, so instead you should change the mode to Logic.

3. Use the data wire to connect the output of your Compare block to the input of your Switch block. The loop should now look like Figure 8.42.

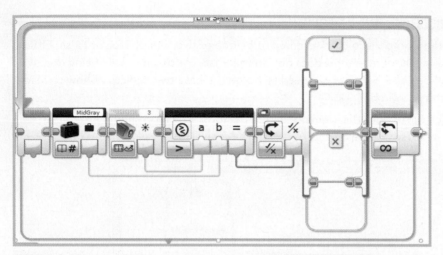

FIGURE 8.42   The data wires are connected for your Switch block.

4. Now you just have to configure the resulting actions. If the Compare block sends the Switch block a True result, you want your bot to go right. If it is False, you want it to go left. To make this happen, drag a Move Steering block into each side of the switch.

5. Set the mode of each Move Steering block to On and the power to 20.

6. Set the angle of the top (True) Move Steering block to 40. It should point to the right.

7. Set the angle of the bottom (False) Move Steering block to -40. The resulting arrow should point to the left. The result should resemble Figure 8.43.

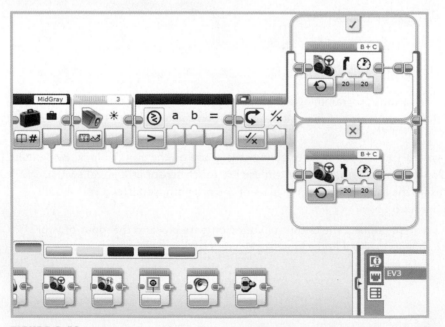

FIGURE 8.43   Insert the Move Steering blocks and adjust as shown.

8. Your program should work now, but let's add one final touch and give it eyes that move back and forth as it steers. Drag Display blocks into both sides of the Switch block.

9. Remember how you found sound files for your Sound block using the Play File setting? You can do the same thing for the Display block. Set the mode to Play File.

10. Click on the text entry area on the top right of the Display block and select LEGO Image Files, Eyes (see Figure 8.44).

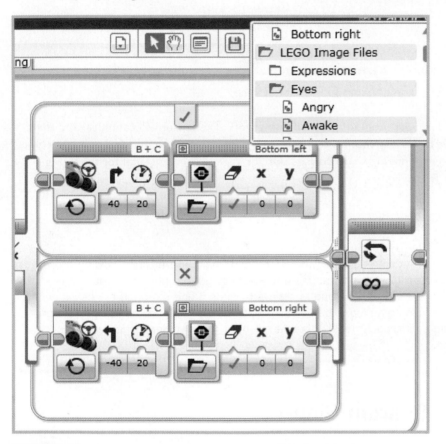

FIGURE 8.44 Find your Image files for the Display blocks.

11. Select Bottom Left for one Display block and Bottom Right for the other.

12. Save your program.

Okay, that should be it. Check to make sure the end of your loop looks like Figure 8.45.

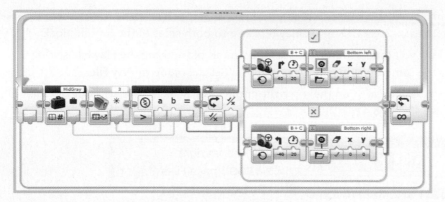

FIGURE 8.45   The line-seeking loop is complete.

Go ahead and test your program. It should now work. The sensor will calibrate, and the robot will move in a zigzag motion as it follows the line. You might notice that it sometimes gets lost when encountering sharp corners or right angles. This is normal for the zigzag method of line following.

FURTHER ADVENTURES

You can experiment with adjusting the speed and angle of the turns, adding more switches for different behaviors, such as a switch to go straight when the line is exactly at the midpoint gray instead of always turning in one direction or the other. You could even add some logic to have the robot turn at sharper angles proportionate to how big a difference there is between what the Color Sensor is seeing and the MidGray value.

Creating Custom Blocks

You can copy and paste blocks and sequences of blocks between other programs, but if there's a sequence that you think you might use often, the most efficient way to preserve it is to turn it into a custom block. Let's make a Countdown block that uses your countdown sequence.

1. Shift and drag your cursor to select the entire countdown sequence (see Figure 8.46)

FIGURE 8.46   The Countdown sequence.

2. In the EV3 software menu, select Tools, My Block Builder.

3. You'll see the My Block Builder menu. Choose a name for your block. It must be all one word and relatively short (see Figure 8.47).

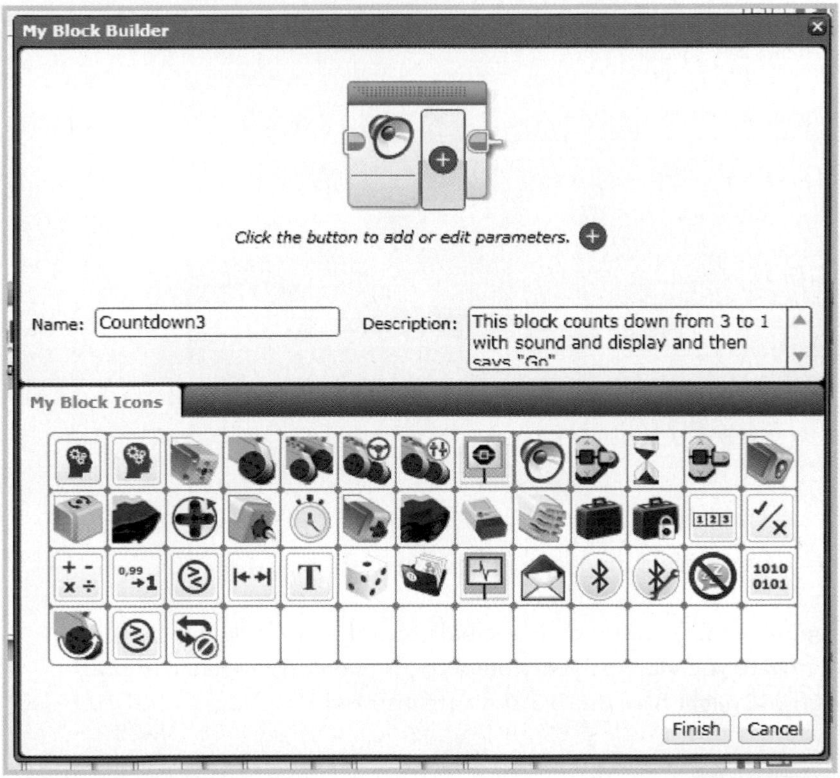

FIGURE 8.47 Add a name and description for your block in the My Block Builder menu.

4. Add a description.

5. Choose an icon.

6. Click Finish when you are done.

After you click Finish, your block is created and your sequence is replaced with the custom block you just made (see Figure 8.48).

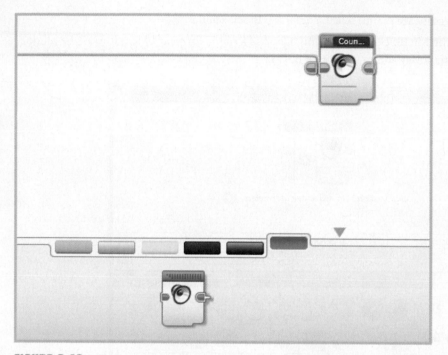

FIGURE 8.48 The custom block appears in the teal tab.

Your new block appears under the teal tab in the palette and is available for you to use in any other EV3 program you make on your computer. (It won't be available to other computers on which you might have the EV3 software installed.)

TIP

If you want to share your custom block with others or other installations of the EV3 software, you can

1. Go to project properties (the wrench on the upper left corner of the screen).

2. Go into the My Blocks tab.

3. Select your custom block.

4. Choose Export.

Documenting Your Work

As you complete this project and others like it, be sure to leave yourself notes about what you've done. That way if something doesn't work or you get an idea on how to change it, you know where to start. Having documentation is also helpful if you want to share the program with others. Figure 8.49 shows some simple notes you can leave using the Comment tool.

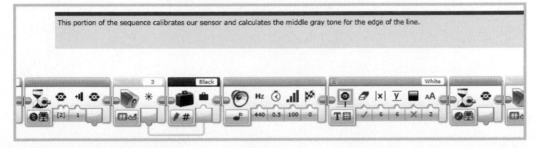

This portion of the sequence calibrates our sensor and calculates the middle gray tone for the edge of the line.

FIGURE 8.49   Leave yourself a few comments.

Summary

In this chapter you dug deep into the heart of making a mid-length program. You made a robot that follows the edge of a black line. It also calibrates the sensor using variables, displays text, and uses audio files. You even created a custom block from a portion of the program you created, which you can now use in other programs.

9

Engineering the Floor-Cleaning Robot

In Chapter 8, "More MINDSTORMS Programming: The Line-Following Robot," you made a line-following robot using the basic robot design from Chapter 7, "Make Your First EV3 Program," and the color sensor. You also used the screen display and EV3 speakers to add the finishing touches. In this chapter, you build upon the skills you learned to construct the line-following robot and make a floor-cleaning robot.

Like the line-following exercise, the floor-cleaning exercise is another classic problem for robotics. Sometimes classrooms ignore the engineering part of this problem and just ask students to program a robot that visits all the spots on a grid. You're not going to do that. Your robot should actually be useful as a floor cleaner. It won't be able to tackle carpets—that requires vacuum parts, but it can clean hard floors.

Before getting to the big project, this chapter familiarizes you with some simple projects that involve programming the infrared and touch sensors, and then you'll take what you've learned about programming and tackle the engineering problems of making a robot that can actually clean floors. Your first mini-project is the self-driven, collision-avoiding robot.

> ### TIP
>
> Be sure to save all your projects. You'll need these programs for your final build in this chapter.

Programming a Collision-Avoiding Robot

Learning to program collision avoidance is a classic programming challenge. Can you make a robot that avoids hitting the walls and other objects without specifically steering it? Yes, you can. Make sure you have a room with a level, flat surface—no carpets. You should also avoid stairs or areas that have sudden drops. This robot does not have any sensors aimed straight down, so it will not detect the drops, and they could potentially damage your EV3.

The first thing you should do to prep for this project is to modify the vehicle you built in Chapter 8. The end result should look similar to Figure 9.1, where the infrared and touch sensors are mounted to the front of the vehicle.

FIGURE 9.1 Your modified vehicle.

Here are the instructions for building this vehicle:

1. Get the vehicle you made in Chapter 8.
2. Remove the color sensor and the beams you used to mount it.
3. Mount the infrared sensor. This should go on the long beam protruding from the front of the vehicle and should be mounted using an axle.
4. Mount the touch sensor to the other front beam using an axle.
5. Connect the infrared sensor to sensor port 3.
6. Connect the touch sensor to sensor port 2.

TIP

If you use the infrared sensor as a remote control, it doesn't matter whether the "eyes" are toward the front of the vehicle or up at the ceiling. However, having the "eyes" toward the front does matter when you're using the sensor for proximity detection.

You're going to do two things with this vehicle: autonomously avoid objects (walls and any toys or furniture), and use the touch sensor as an on and off button. Using what you learned in Chapter 7 and Chapter 8, you're going to use loops and a switch to have the bot detect whether an object is nearby and, if something is, then to steer away from it.

Activating the Touch Sensor

You want to be able to activate your robot by using the touch sensor. Here's how you create that portion of the program:

1. Open the EV3 Home Edition software and start a new project.

2. Drag a Wait block onto the canvas next to the Start block.

3. Change the mode of the Wait block to Touch Sensor, Change, State, as shown in Figure 9.2.

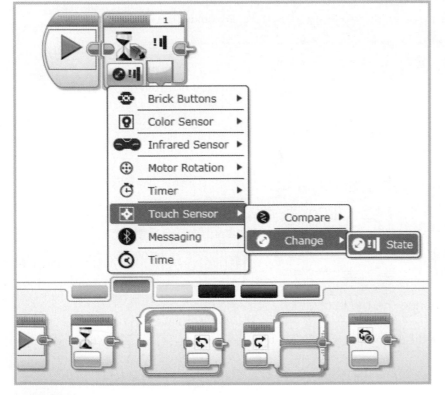

FIGURE 9.2   Change the mode of your Wait block.

That's it. The Wait block's lone function is to wait for something to change before it allows the program to proceed to the next block. In this case, you've specified that any change in the touch sensor's state triggers the next action.

NOTE

The touch sensor actually has three states: pressed, released, and bumped. *Bumped* means that the touch sensor has been pressed and then released. In other words, you won't accidentally trigger two events when briefly pressing the button. You could use a touch sensor in front of a vehicle as a collision detector, although it would detect the collision after it had happened. The infrared sensor detects potential collisions before they occur.

Adding Collision Avoidance

Now that you've programmed the touch sensor, the next step is to add collision avoidance.

1. Drag a Switch block onto the canvas after the Wait block.

2. Change the mode to Infrared Sensor, Compare, Proximity (see Figure 9.3).

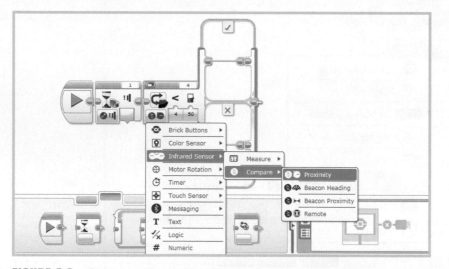

FIGURE 9.3   Change the mode of your Switch block.

3. Now, change the settings on your Switch block to less than (<) 5 as shown in Figure 9.4.

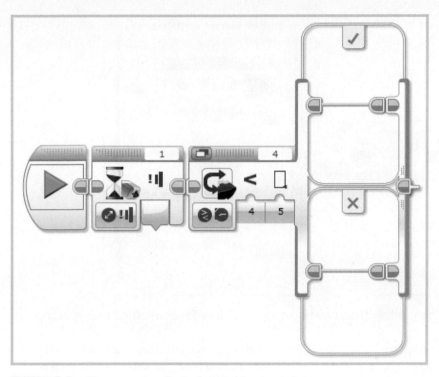

FIGURE 9.4 Change the Switch block's settings.

At this stage, the infrared sensor's proximity mode tries to detect distance from objects in a range between 0 and 100, with 100 being very far away from anything and 0 being very, very close. It does this by sending out an infrared signal and measuring how long it takes the signal to bounce back to the sensor. It's not 100% accurate. The surface the signal is bouncing against could diffuse the signal and confuse the sensor, for example. For that reason, you'll give the sensor a bit of leeway and specify that the number has to be under 5. That proximity is close, but it isn't right on top of an object.

Now you have to decide what happens when the value returned by the infrared sensor is true or false.

Have the robot turn if the value is True (the infrared sensor detects a nearby object) and go straight if the value is False. Add some Move Steering blocks. Drag the Move Steering blocks into place, and give the True block a turn radius of –56 and the False block a value of 0 (see Figure 9.5).

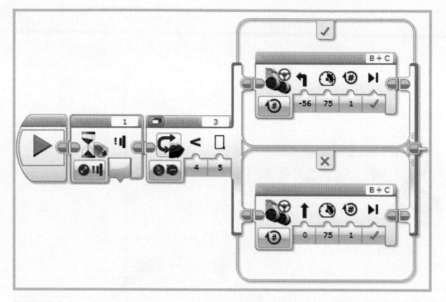

FIGURE 9.5 Change the turn radius to –56 for the True value of this switch.

Right now, the action you've set up with this switch only happens once, and you want everything to keep going, so put the switch inside a Loop block, as shown in Figure 9.6.

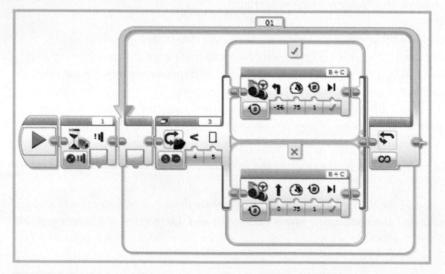

FIGURE 9.6 The Switch block is now inside the loop, but the Wait block is not.

Testing Your Bot

Go ahead and test your program by running it on your EV3. Remember to start the vehicle by pushing the button on the touch sensor.

How does it work? This is a trick question. If you built your bot exactly the way presented in this chapter, nothing will happen. That's right—your bot won't go forward, sideways, or in any direction.

Why? If you look closely, you'll see that the program is paused at the first Wait block. The port of that Wait block is set to 1. Your Wait block is actually plugged into port 2. So, before you go any further, make sure your Wait block and Switch block are set to the correct ports, as shown in Figure 9.7. The motor ports should already be correctly set to B and C.

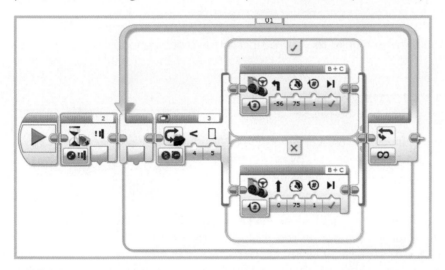

FIGURE 9.7 Always check your sensor and motor ports!

As a programmer, you must constantly keep a watchful eye on these kinds of details. Overlooking small details that will, in turn, have a huge impact on your bot's ability to function is incredibly easy to do!

Now that you've run into trouble, this a good time to add some bug tracking to your program. Let's change the screen depending on whether the robot is turning or going straight. That way you can tell whether the infrared sensor is actually working the way you expect it to work. Follow these steps:

1. Add Display blocks to each side of your Switch block.

2. Change the False value (no proximity detected) to the "big smile" image.

3. Change the True value (close to an object) to the "sick" image. You can see this in action in Figure 9.8.

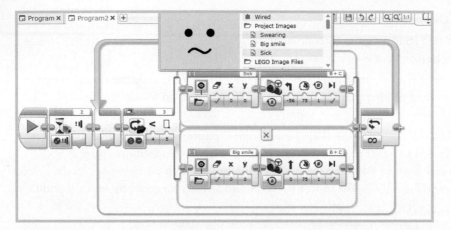

FIGURE 9.8   Add Display blocks to your program for troubleshooting.

TIP

Changing your screen display or adding noise not only makes your program seem more polished, but it also helps you. By adding these elements, you can troubleshoot your program and verify that it is working as designed.

Now, go ahead and test your program by running it on your EV3. If it is working as designed, it will wait for you to press the touch sensor button and then start going straight. It will then turn when it gets near an object.

Navigating Corners

You might notice that your robot sometimes ends up getting stuck in corners. Corners are a bit tricky for the infrared sensor to navigate because turning doesn't always get the robot away from the obstacle.

To fix this problem, make the robot back up a little each time it has to turn. Add another Move Steering block to the "true" side of the Switch block, and change the motor power value to –53 (see Figure 9.9). This makes the robot back up a bit before turning, and it should prevent it from getting stuck in most corners.

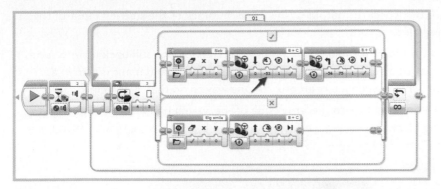

FIGURE 9.9   Add another Move Steering block to back up your bot.

Adding a Bit of Randomness

Wait—don't robotic floor cleaners actually do more than go in a straight line and turn in a single direction? What you need, if you want to keep this robot scooting along, is an element of randomness. Fortunately, the programming palette contains exactly what you need:

1. Add a Random block onto the "false" side of the Switch block, as shown in Figure 9.10.

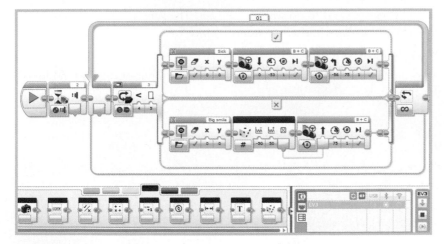

FIGURE 9.10   The Random block looks like a six-sided die.

2. The Random block can generate either a number or logic value (Yes or No). In this case, you want a number. Pick a number between –50 and 50. You can tweak it later if you don't like the results.

3. Set the inputs of the Random block to –50 and 50.

4. Use a thread to connect the output of the Random block to the steering input of the Move Steering block.

Okay, now test your robot. It should go in a random direction but still back up and turn when it encounters an object. You could take this a step further and make it turn in a random direction when it encounters an object. One way to do that is as follows:

1. Add another Random block to the "true" side of the Switch block. Place it after the Move Steering block that moves the robot backward but before the block that turns it, as shown in Figure 9.11.

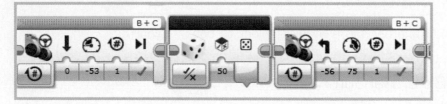

FIGURE 9.11   Insert a Random block.

2. In this case, set the Random block mode to Logic. Notice how this changes the input to something that currently says "50." That's your probability setting. Yes or no both have a 50-50 chance of occurring. The default setting works for our purposes, so leave it as is.

3. Now you have to figure out what to do with the output. The simplest approach is to add another Switch block. Place it after the Random block.

4. Change the mode to Logic.

5. Use a thread to connect the output of the Logic block to the input of the Switch block (see Figure 9.12).

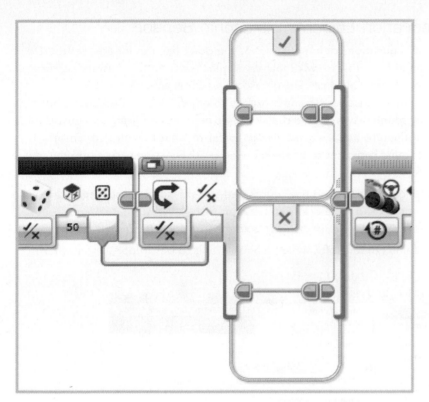

FIGURE 9.12   The Random block now controls the Switch block.

What this does is make a random choice for the switch. Now you can move the Move Steering block into one side and add a Move Steering block turning in the opposite direction to the other side. The final program should look something like Figure 9.13.

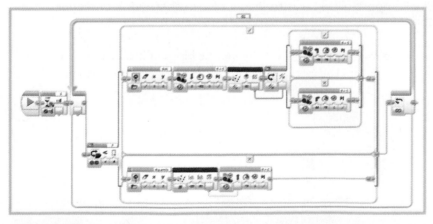

FIGURE 9.13   The final collision-avoiding program.

Using the Education Edition's Ultrasonic Sensor

If you have the LEGO Education edition of the EV3, you don't have an infrared sensor. Do not fret. Your version of this program will look very similar but use the ultrasonic sensor. Think of it as your own personal sonar. The ultrasonic sensor is actually much more accurate than the infrared sensor at sensing proximity to objects and walls. It sends out super high-frequency sound waves (higher than what we humans can hear) and measures how long it takes for them to bounce back. This is similar to what the infrared sensor does with light, but sound turns out to be the more accurate way to measure proximity to objects.

If you have the EV3 Home Edition and have purchased the ultrasonic sensor separately, you can also use it. First download the block from LEGO at http://www.lego.com/en-us/ mindstorms/downloads/ev3-blocks/ultrasonic/, and then follow these instructions:

1. Go to the EV3 Home Edition software and choose Tools, Block Import, as shown in Figure 9.14.

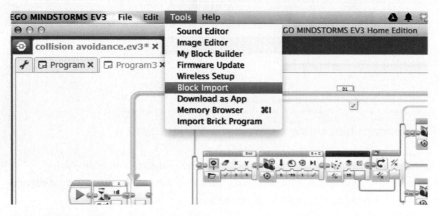

FIGURE 9.14 Block Import allows you to install blocks that didn't come with the default installation.

2. A window appears, allowing you to browse to find the block, if it isn't already selected by default (see Figure 9.15). Locate the Ultrasonic.ev3b block and click Import.

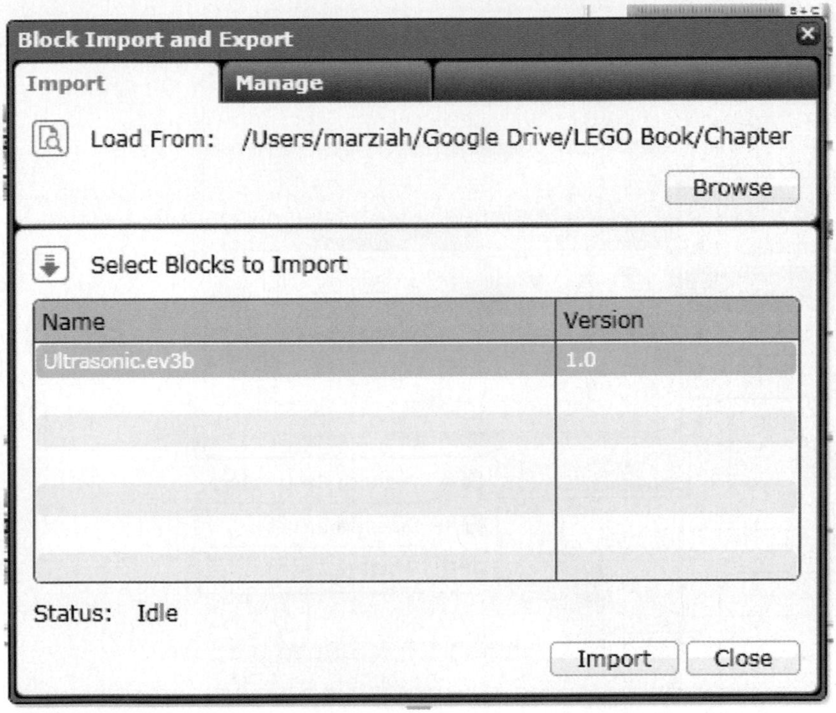

Block Import and Export   ✕

Import	Manage	

◱ Load From: /Users/marziah/Google Drive/LEGO Book/Chapter

Browse

⬇ Select Blocks to Import

Name	Version
Ultrasonic.ev3b	1.0

Status: Idle

Import   Close

FIGURE 9.15   Browse to find your downloaded block and click Import.

3. A warning appears, telling you that you must restart the software for the change to take effect. After you do so, you'll see your new block in the palette as if it had always belonged there (see Figure 9.16).

FIGURE 9.16   You now have the Ultrasonic sensor block.

Now that you have access to the functions of the ultrasonic sensor, you can just make one little change to your program. In the initial Switch block, change the sensor from infrared to ultrasonic. You'll notice it offers different choices, as shown in Figure 9.17.

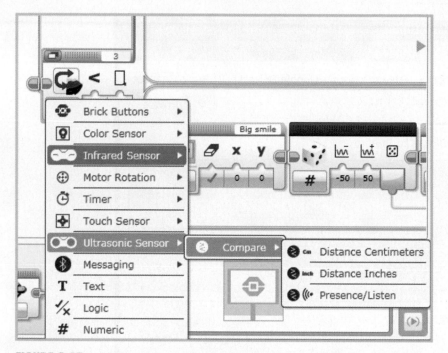

FIGURE 9.17   Here are the Ultrasonic sensor options.

Rather than giving proximity as a number between 0 and 100, the ultrasonic sensor can give you that number in inches or centimeters. Go ahead and pick centimeters, and change the input from "5" to "2." Your bot might as well get as close to the wall as it can without touching it, and the ultrasonic sensor makes it easier for that to happen.

Unfortunately, there is no ultrasonic equivalent for the infrared remote, which is what you'll program next, but the good news is that your self-guided robot does a better job at collision avoidance than a robot that relies only on an infrared sensor.

Controlling Your Bot with the Infrared Remote

Now that you've (mostly) avoided collisions in your self-guided robot, it's time to take things into your own hands. Let's use the infrared sensor beacon to make this a remote-controlled robot.

First, take a closer look at the infrared beacon and sensor, as shown in Figure 9.18.

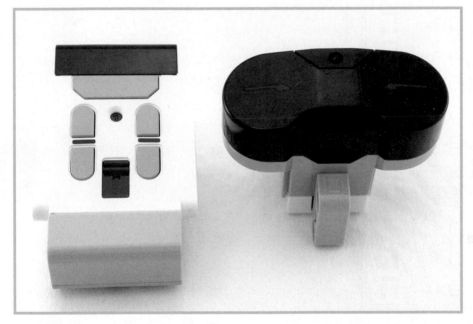

FIGURE 9.18 Control your bot with the infrared beacon/remote and sensor.

The beacon or remote (same device) works a lot like the remote controls for many TVs. It sends an infrared signal to the infrared sensor, which, when hooked up to a sensor port on the EV3, can be used in programs to trigger commands.

The remote has five buttons, but there are actually twelve possible states for the buttons. No buttons pressed, one button pressed, and any combination of two of the smaller buttons pressed (see Figure 9.19).

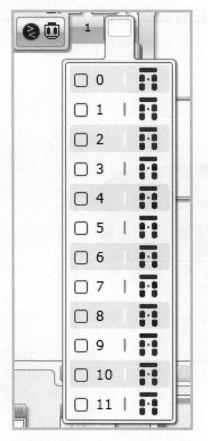

FIGURE 9.19 You can see all the possible button states on the remote.

All these button states offer you a lot of programming flexibility, but they also mean you need to map them out carefully to figure out exactly what you want the remote to have your bot do. Let's keep it simple with forward, left, right, and backward. Figure 9.19 shows the map for the remote.

Programming the Remote

One way to program your remote is to create a series of switches to go through each of your buttons and determine whether to take an action. This is the way LEGO chose to program the remote for the R3ptar and many of the other demo robots (see Chapter 4, "Building Your First Bots," for more info).

1. Open the EV3 Home Edition software and start a new project.

2. Drag a Switch block onto the canvas.

3. Change the mode to Infrared Sensor, Compare, Remote, as shown in Figure 9.20.

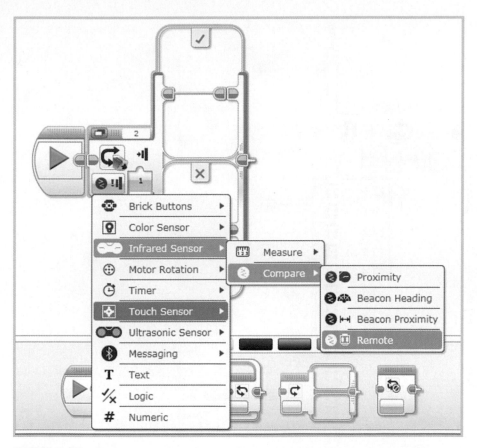

FIGURE 9.20 Set the mode of your Switch block to use the remote.

4. Change your port to 3, because that's where the infrared sensor is plugged in.

5. Change the input of your Switch block to use the front button on the remote. This is position 9, as shown in Figure 9.21.

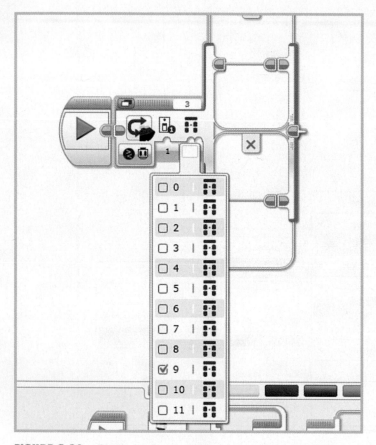

FIGURE 9.21   Change the input of your Switch block to use the front button.

NOTE

Notice that you can check more than one box at the same time. If you do that, any of the checked boxes will make a "true" event on the Switch block.

Now let's add a "true" action. In this case, the Switch block is detecting the top button, which is the forward button, so drag a Move Steering block onto the "true" side of the Switch block, as shown in Figure 9.22.

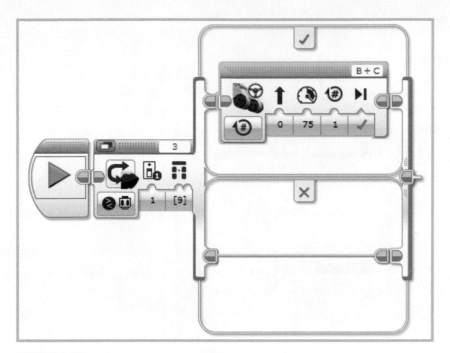

FIGURE 9.22   Your program so far.

What goes in the "false" side of this switch? Why, another switch:

1. Drag a Switch block onto the "false" side of the first Switch block, as shown in Figure 9.23.

2. Adjust the settings to have this block triggered by the left button on the remote, input 1.

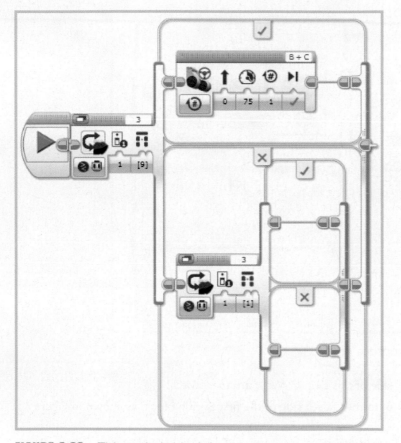

FIGURE 9.23 This switch block is set to react to the left button on the remote.

3. Place a Move Steering block on the "true" side, and adjust the inputs to make it turn the robot left.

4. Add another switch to the "false" side.

5. Keep going until you have four switches and four different button actions to control the directions of forwards, right, left, and backwards.

6. Drag everything into a Loop block.

Your final program should resemble Figure 9.24.

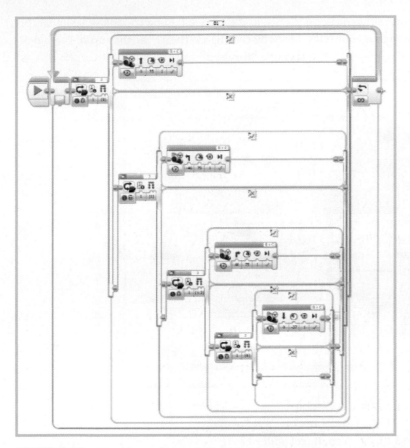

FIGURE 9.24 The four nested Switch blocks program the remote to respond to button presses.

Go ahead and test the program. It should work. However, it seems like there should be a better way to get this program to work. Don't worry, there is: Use a multi-threaded remote program.

Creating Multi-threaded Programs

The EV3 programming language is capable of understanding more than one program running at the same time, as long as they're not all using the same resources at the same time. You can't make a motor go both forward and backward at the same time, for example. You can, however, have your robot wait for left, right, up, or down button presses. Here's how:

1. Start a new project in the EV3 Home Edition software.

2. Drag a Loop block onto the canvas.

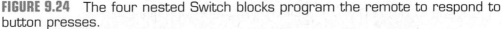

3. This project uses four different Loop blocks, and they're all going to look very similar. So to avoid confusion, name your Loop block by clicking on the "1" at the top of the block and typing in a new name. Let's call this one **"forward"** as shown in Figure 9.25.

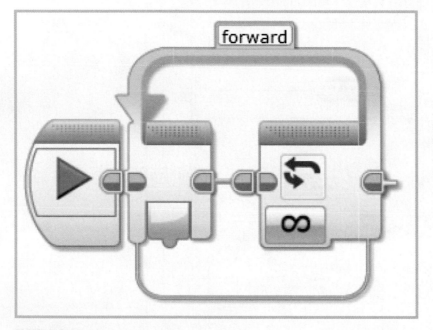

FIGURE 9.25   Rename your Loop block.

4. Add a Wait block to the loop to detect your remote button.
5. Change the mode of the Wait block to Infrared Sensor, Compare, Remote.
6. Change the input to the top button [9]. (Be sure to adjust your port settings if necessary.)
7. Drag a Move Steering block into the loop after the Wait block. You should have a block that looks like Figure 9.26.

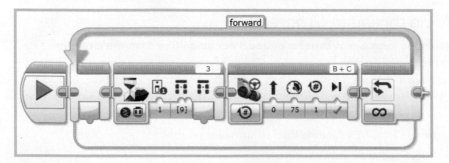

FIGURE 9.26   The completed forward loop.

NOTE

This program is super simple, and it will work if you want to test it. It just makes the robot go forward when you press the remote button.

8. To make this a multi-threaded program, drag another Start block onto the canvas, as shown in Figure 9.27.

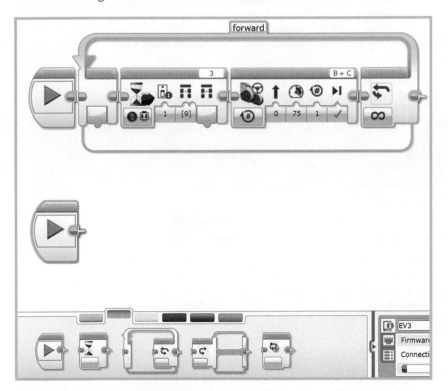

FIGURE 9.27 Drag another Start block onto the canvas.

9. Repeat the same basic steps as you did for the first program, except name this Loop block **"left,"** and change the input of the Wait block and Move Steering block to make this loop move the robot left when the left top button is pressed on your remote.

After you've completed those steps, you should have something that looks like Figure 9.28.

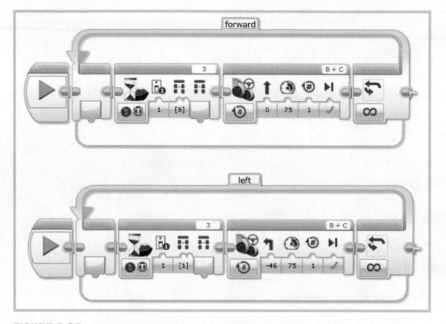

FIGURE 9.28   Two loops with two different Start blocks.

10. To complete the program, add the threads that will make the bot go right and back and rename them accordingly. You should have four loops total, as shown in Figure 9.29.

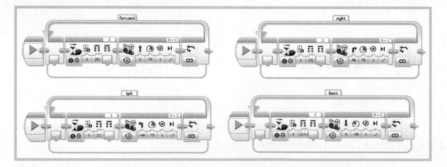

FIGURE 9.29   You can see all four threads in the remote program.

Go ahead and try running this program now. It should work just as well as the other more complicated nested Switch blocks version of the program.

If you want to take the program a step further, you can make your remote a little more user friendly by accounting for a common error. Your remote requires you to press both bottom buttons to make the bot go backwards, but you're not using a single bottom button press for any other function, so why not add that possibility to your remote? That way if someone isn't pressing quite hard enough on the two buttons, it will still go backwards as they intend.

To add this fix, open the input on the Wait block in your "Back" loop. Select the 2, 4, and 8 options as shown in Figure 9.30. You can have them all checked at once, and then the infrared sensor will react to any one of them (as opposed to all of them at once).

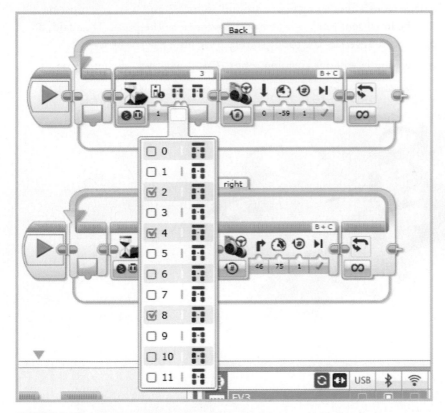

FIGURE 9.30 Select multiple options for your remote.

Adding the Floor-Cleaning Functions

Here's your chance to put all the ideas together and make a robot that will really clean your hard floors.

What you need:

- EV3 Home Edition (you're going to modify your current vehicle build, so don't tear it down just yet)
- A sponge mop replacement head (available from Target or similar stores)
- Your previous programs

You might need to modify the upcoming instructions to meet with your particular replacement mop. I chose a wide cellulose sponge mop-head replacement with a firm plastic backing. You could also modify this project to work with narrower cellulose sponges, microfiber towels, or other materials. The point is to have a slightly damp cleaning cloth or sponge that your robot can drag across a hard surface floor.

You might be a little nervous about using water around your EV3 and, yes, you must be careful not to dunk the Intelligent Brick in water or get water into the ports. That said, a slightly damp mop being dragged by the robot should pose little risk to the EV3.

Figure 9.31 shows the completed floor-cleaning robot. This is what you're aiming for.

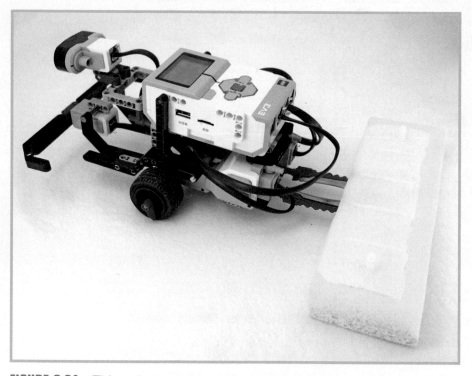

FIGURE 9.31   This robot uses a replacement sponge mop head to clean your floor.

Building the Mophead Assembly

Now that you have seen the solution, let's take a look at how you get from the bot pictured way back in Figure 9.1 to the one shown in Figure 9.31. There are only two wheels, so a little rear weight goes on to the mop head to help drag it around with good contact on the floor surface. The front has a "bumper" made from a touch sensor, and the infrared sensor is used for collision avoidance. Modify your current vehicle build as follows:

1. Remove the caster wheel and frame on which it is supported. The rear caster wheel served the vehicle well for the previous exercises, but now it would prevent the rear of the robot from getting enough weight onto that mop head.

   Instead, use two EV3 "swords" to secure the sponge mop head to the rest of the vehicle. Figure 9.32 shows the basic parts: two pegs, two 2M pegs with cross holes, two swords, and a 9M beam.

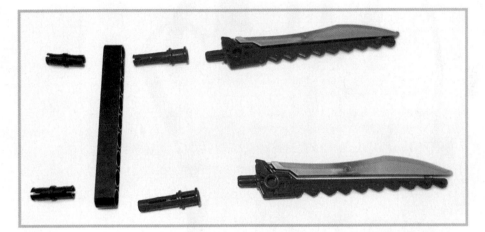

FIGURE 9.32 The basic parts for the sponge mop head attachment.

2. Insert the black pegs into the back of the large motor servos, as shown in Figure 9.33.

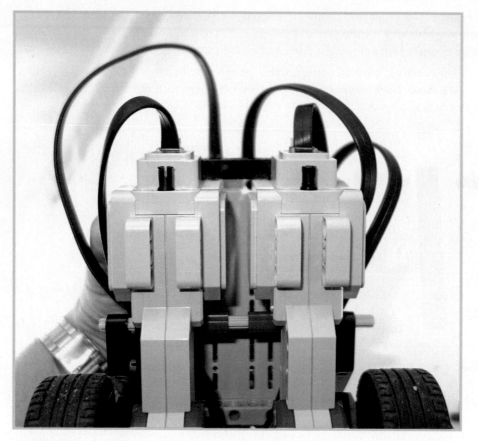

FIGURE 9.33   Insert the black pegs into the servos.

3. Use the pegs to secure the 9M beam. This adds stability to the rear of your vehicle.
4. Add the red 2M pegs with cross holes (see Figure 9.34).

FIGURE 9.34 Notice how the red 2M pegs with cross holes are both aligned with the inside hole of the servo motor.

5. Add the swords into the cross holes, as shown in Figure 9.35.

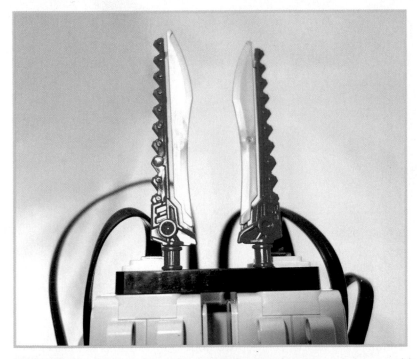

FIGURE 9.35 Notice how the red edges point toward the outside edges.

It's possible you could jab the swords right into the sponge at this point, but that might also be a recipe for breaking a few LEGO pieces. A safer method is to line up the sponge and press the swords just firmly enough against the sponge to make an impression. I then took scissors and jabbed holes into the sponge just wide enough to allow the LEGO swords to fit into the sponge.

Adjusting the Sensor Assembly

Now that you've assembled the back, let's talk about the front. The front of the floor-cleaning robot uses a belt-and-suspenders approach where you employ the infrared sensor to avoid objects and the touch sensor "bumper" to back up in case a collision happens anyway. Figure 9.36 shows the bumper assembly.

FIGURE 9.36 The bumper assembly.

Follow these steps to make this assembly:

1. Join two double-angle beams to a size 15M beam using four black pegs. Leave three beam holes free in the center.

2. Using blue connectors, fasten a red double-cross block.

3. Switching your attention to the vehicle, connect two double-angle beams to the inside of the double angle beams currently on the vehicle. Use the gray 3M special connectors.

4. Fasten the touch sensor to the bumper assembly using an axle through the double-angle beams.

5. Attach the bumper portion to the touch sensor using another blue peg cross-axle, as shown in Figure 9.37.

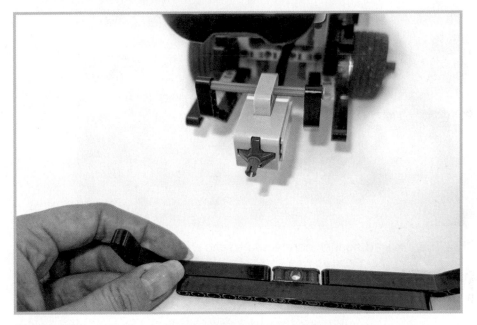

FIGURE 9.37 Connect the bumper to the touch sensor.

Essentially what the bumper is doing is making a bigger surface for your touch sensor to bump if it collides with anything. Because it is connected by a pin and not by an axle, it will turn if you move it, but the pin you used has some friction, so it should also be relatively stable.

> **NOTE**
>
> Be sure to connect the touch sensor to sensor port 2 if you have not done so already.

Now you need to install the infrared sensor. Let's put it above the bumper and centered on the front of the vehicle. Figure 9.38 shows the assembled sensor.

FIGURE 9.38   The infrared sensor connected to the front of the vehicle.

Follow these steps:

1. Connect the infrared sensor to two 3M cross blocks as shown in Figure 9.39.

FIGURE 9.39   Attach the sensor using cross blocks.

2. Attach the cross blocks to a beam frame using black pegs.

3. Attach the beam frame to the robot. I used angle beams and 2M pegs with cross holes. Figure 9.40 shows the build from another angle.

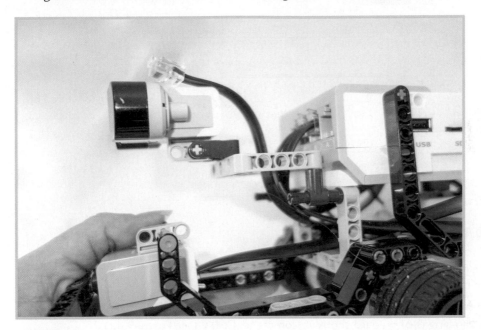

FIGURE 9.40 Don't forget to connect the cable to the sensor and sensor port 3.

Building the Floor Cleaning Program

Now that you've completed the engineering task, it's time for the programming. The good news is that you've already done most of the work by following along with the earlier chapters.

1. Open the collision avoidance program you created at the beginning of this chapter (refer to Figure 9.13).

2. Click and drag your mouse to select the entire program.

3. Click on Edit, Copy as shown in Figure 9.41 (you can also just use the keyboard shortcut Ctrl+C on Windows or Command-C on Macintosh).

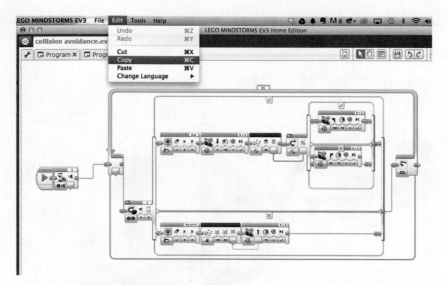

FIGURE 9.41 Copy the entire program.

4. Open a new project. You can keep this project open at the same time or close it. It doesn't matter.

5. Paste the old program into your new project window. Do this by either selecting Edit, Paste or using the keyboard shortcut Command/Ctrl+V (see Figure 9.42).

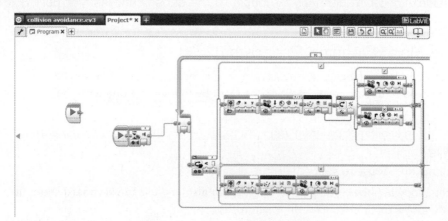

FIGURE 9.42 Paste the program into a new project.

You now have two Start blocks, but that's okay. You're going to have two threads in this program. Otherwise, you could just delete the extra Start block.

6. Click on the canvas to deselect your entire program.

7. Click to select the Wait block just after the Start block and delete it. The program will be disconnected from the Start block and appear a little lighter to indicate that it isn't currently functional.

8. Connect the rest of the program back to the Start block using a thread or just by dragging the Start block next to the Loop block (see Figure 9.43).

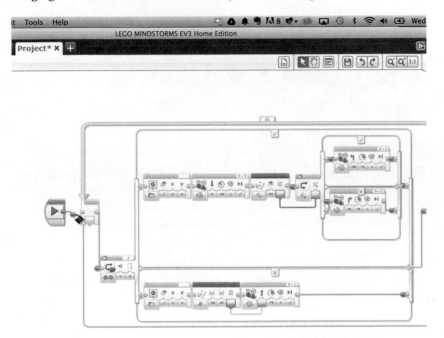

FIGURE 9.43   Connect the Start block to the Loop block.

After everything is connected, the whole program should look dark again.

Now, because you already wrote this program earlier in the chapter, tested it, and connected the sensors to the same ports, you're ahead of the game. Unlike the first program, this won't wait for a touch sensor bump to start working. It will immediately run. However, you've done nothing for the touch sensor. Let's remedy that by using the spare Start block already on your canvas.

This time you'll make another small loop like you did with your remote control program earlier in this chapter, only this time use the touch sensor to make the robot back up any time it runs into an object:

1. Drag a Loop block next to the Start block.

2. Drag a Wait block into the loop.

3. Change the mode to Touch Sensor, Change, State.

4. Drag a Move Steering block after the Wait block.

5. Change the inputs to –53 power and 2 rotations.

6. Rename your loop **"touch sensor"** to remind you why it is in the program.

The completed loop appears in Figure 9.44.

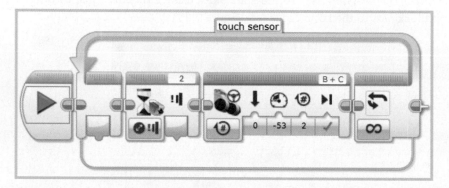

FIGURE 9.44 The simple loop backs up the robot when it runs into something.

Go ahead and take your robot for a test drive. How did it do? It should make slight random turns and go generally forward while avoiding objects and backing up when it runs into something. That's actually similar to how some high-priced floor cleaning robots work, though I'll grant you they're much more sophisticated about it.

If you're using the LEGO Education version, you can substitute an ultrasonic sensor for the infrared sensor as you did earlier in the chapter and follow all the other instructions.

> **TIP**
>
> If you want a remote control floor cleaner, you've already made the program to control the remote in Programming the Remote in this chapter. It should work without any modification.

Summary

In this chapter, you modified the bot you built in Chapter 8 by programming it to use the infrared and touch sensors to avoid collisions. You learned a way to help your bot navigate corners, and you learned how to enable your bot to move randomly. You discovered the usefulness of multi-threaded programs, and you engineered a mop-head assembly for your bot and built a program that turns it into a useful floor-cleaner. In the next chapter you'll learn how to make a robot deal a deck of cards and identify them by color.

10

The Color Magic Card Trick

In this chapter, you'll make a robot that identifies cards by color and deals them out one at a time. Your robot can tell you the color of the card, even when you can't see it. No, it's not magic, but it's still pretty cool.

In addition to the EV3, you need a deck of Uno playing cards, preferably an older one. The older editions of Uno have solid colored backgrounds, whereas the more recent editions have mottled rays of color, which make it harder for the color sensor to detect the color correctly. If you have one of these newer decks, don't worry—we'll go through testing the deck to see whether you need to take some colors out of your deck for the trick to work correctly. As an alternative, you can also use Skip-Bo or Phase 10 cards. All three are made by Mattel and available in most retail stores.

This project offers you two challenges:

- To engineer a robot that deals cards
- To program the robot to identify the color of the cards, face down

Figure 10.1 shows the project flowchart.

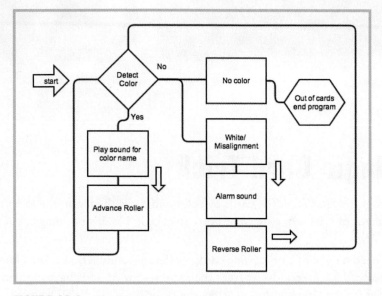

FIGURE 10.1   The flowchart could be larger if you wanted to list each detected color individually.

Brainstorming and Building the Bot

Engineering a project such as this one takes a lot of trial and error, so brainstorm some ideas and try out some models. You can try out the mechanics of ideas without having to attach the motors at this stage. Instead of using the motors, you can just physically turn the piece that would be powered by the motor to test out an idea.

The robot has to deal cards, so how would that work? You could engineer some sort of hand that grabs cards from the deck. A gripping hand might require gears to open and close and some way to grab just one card instead of the whole pile. Originally, I thought about something along the lines of the LEGO Education robot in Chapter 5, "Building the LEGO Education Bots" (see Figure 10.2). You can try playing with this idea, but I'll tell you now that I am unable to make this work. My big sticking point is getting the hand to grab only one card and still be able to move it from one pile to another.

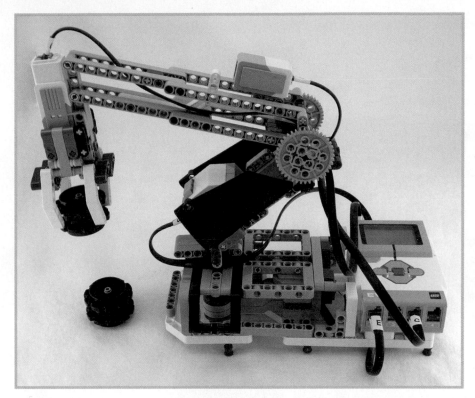

FIGURE 10.2 This robot has a grabbing hand, but it needs large objects to grab. It cannot grab a single card from a pile.

So rather than thinking "hands," think about anything that would pull cards. You could, for example, use the rubber tank track to pull cards from underneath the pile. If the deck were slightly tilted, maybe gravity would help keep the other cards from being dragged around (see Figure 10.3).

FIGURE 10.3 The tank track method isn't going to work out.

In practice, the tank track ends up not working. Even with the aid of gravity, it ends up pulling lots of cards from the deck.

So, what about using a wheel instead of a tank track? The rubber wheel cover should still grip the card well enough to pull it against gravity, and maybe we can adjust it to pull out one card at a time, instead of the whole deck (see Figure 10.4). Since the color sensor detects color from reflected light, wheels would also give the color sensor just enough room to detect the color of the card.

FIGURE 10.4 Test the wheel method by twisting the wheel manually. There's no need to build something complex until you have a mechanism that will work.

After trying to just spin a wheel by hand under a frame beam, I think this idea will work best. When the wheels are doubled up, they're almost the same width as the card deck. If the deck is held at an angle and the wheel actually works against gravity, it might help keep the remaining cards in place.

Building the Platform

Now that we have a basic idea of how this will work mechanically, let's build this robot, beginning with the base platform that holds the cards.

TIP

In the process of building a robot, you may want to experiment a lot. Try rapid prototypes. Adjust things that don't work. Improve things that do work to make them work even better. These steps are the result of building and tearing apart many robot versions.

1. Build a platform for your EV3 Intelligent Brick and your deck of cards. Use three beam frames for a flat, open area that is stable and can accommodate the sensor and wheels we'll add later. Hold the beam frames together with black pegs.

2. Place one 15M black beam across the back edge of the platform for stability. The base should look like Figure 10.5.

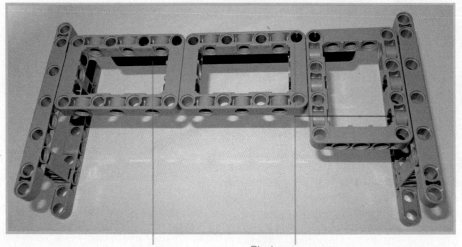

15M beam   Black pegs secure the beams

FIGURE 10.5   A simple platform base holds both the Brick and your deck of cards.

This base is not actually tall enough for the color sensor to fit underneath it, pointing upwards, so you must build in the height later. For now, let's continue getting all the pieces in place.

3. Use your cards as a measuring guide and place beams around three sides of it, with one side running along the right side of the platform to hold the cards in place (see Figure 10.6). Use black pegs to hold these beams in place. Also, add a 90-degree beam and a cross-axle on the lower-right side to eventually hold the wheel axle in place.

FIGURE 10.6   Use the card itself as your measuring guide.

It may be difficult to see where the actual pieces are assembled with the card in the way, so Figure 10.7 provides an unobstructed view. In addition to the straight beams, there is one 90-degree angled beam. This was actually a later addition after several iterations, because it moved the cards into better alignment with the color sensor.

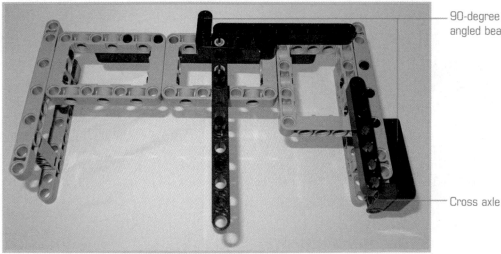

FIGURE 10.7   The card-sorting robot so far.

4. Prepare the color sensor for mounting by attaching two cross beams with a 3M axle and a blue double-wide pin, as shown in Figure 10.8.

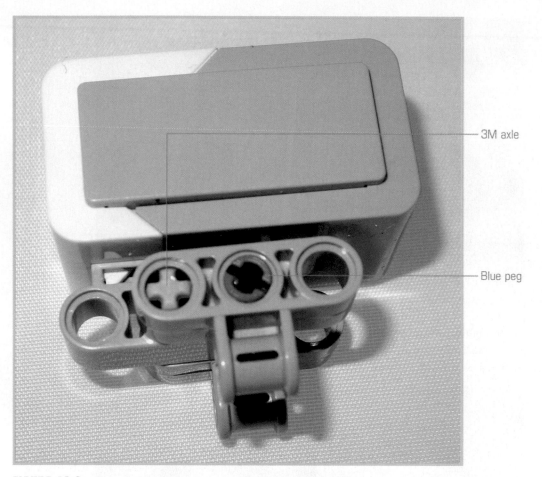

3M axle

Blue peg

FIGURE 10.8 The color sensor is prepared for mounting.

5. Mount the color sensor on the right side of the platform, flush with the frame beam. Use black pins in the cross beams and firmly attach the sensor to the underside of the frame beam that supports the side of the platform (see Figure 10.9). The sensor should be pointed upward, since it will read the color from the card on the bottom of the deck.

FIGURE 10.9   Mount the color sensor by pinning the cross beams to the frame beam supporting the side.

Raising the Platform

Because you're not building a vehicle, you need to lift the platform frame up higher to give the wheels clearance to spin without moving the platform. You also want to tilt the platform at a slight angle, so that gravity keeps the bulk of the cards in place while dealing one. You also want to make sure there's still a lot of contact with the ground on the front of the platform, because tilting the platform back could introduce some instability, and you're going to be adding a heavy Intelligent Brick to the top.

Construct a sturdy frame using one black 15M beam, a red 3M cross-axle beam or standard beam, two double-bent 45-degree beams, two 3M beams, and two 11m red beams (see Figure 10.10).

> **NOTE**
>
> If you use a cross-axle beam for this step, secure it with blue half-axle pins. If you use a standard beam, secure it with black pins.

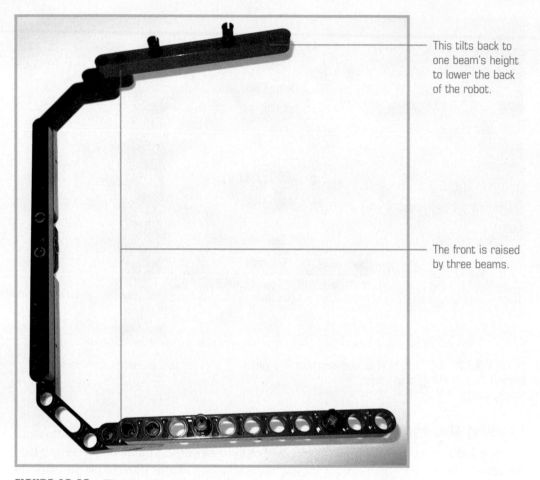

This tilts back to one beam's height to lower the back of the robot.

The front is raised by three beams.

FIGURE 10.10   The beams here raise the platform at an angle.

The end result makes the front of the platform three beams taller, and the end of the platform one beam taller. After installing the "platform shoes" on your platform base, you should have something that resembles Figure 10.11.

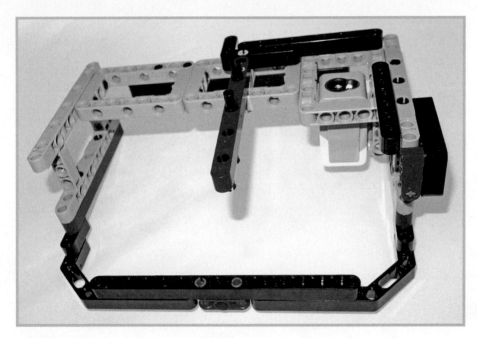

FIGURE 10.11   Our robot now has "platform shoes."

Building the Wheel Assembly

To make the wheel assembly, cover two large wheels with treads, and connect them together through the center of each wheel using the longest axle in your EV3 kit. Place a yellow half-bushing on one end to keep the tires in place. Place the opposite end inside the medium motor, as shown in Figure 10.12

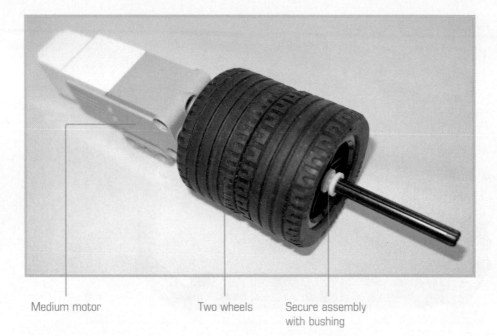

Medium motor   Two wheels   Secure assembly
                                                  with bushing

FIGURE 10.12   The wheels connected to the medium motor.

At this point, the wheel assembly needs to go underneath the area where the cards will rest. Attach the medium motor to the middle beam, and place the remaining length of axle through the red cross-axle beam and black 90-degree beam on the right side of the card deck area. The top of the wheels should be just slightly higher than the top of the color sensor. Secure the axle with another yellow half-bushing. The assembly should now look similar to Figure 10.13.

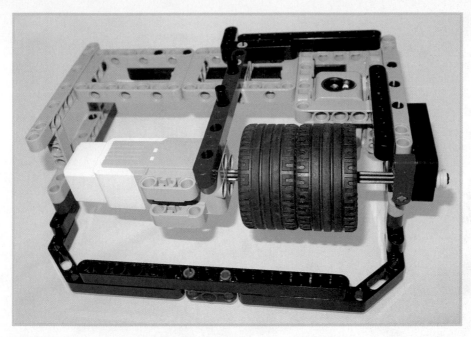

FIGURE 10.13   The medium motor and wheels are in place.

Checking the Assembly

At this point, verify that everything is correctly positioned. Place your cards on the bot to verify that the color sensor is correctly positioned to detect the solid color in the upper-right corner of your card (see Figure 10.14). You should also double-check that the cards are leaning at a slight angle toward the back of the bot (this keeps the entire deck from sliding forward every time the wheel rotates). Finally, try rotating the wheels by hand to make sure it still pulls cards outward.

If the assembly does not work as you anticipated, here is your chance to problem-solve. You can try leveling the platform or raising the wheel height to see if you can get the cards to move correctly.

FIGURE 10.14   The deck should be resting at a slightly backward tilt.

Placing the Intelligent Brick

Now it is time to place the Intelligent Brick on the bot and connect the cables to the sensor and motor. Secure the Intelligent Brick to the left side of the bot using black pins (see Figure 10.15).

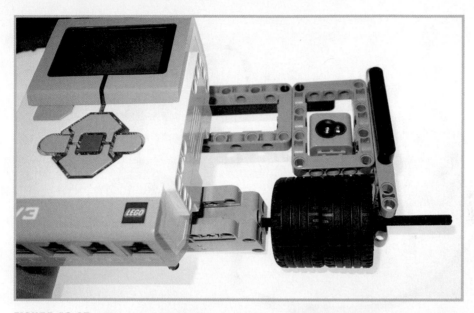

FIGURE 10.15 The Intelligent Brick goes on the left. In this photo, I removed the card deck area to better show the placement.

Flipping the deck over may be easier for attaching the cables. Connect the medium motor to port C, and connect the color sensor to port 3, as shown in Figure 10.16.

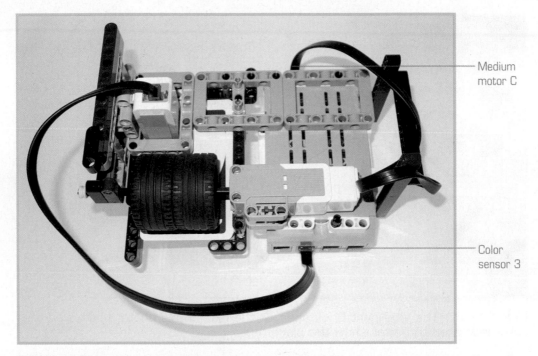

FIGURE 10.16   Here is the underbelly of the bot (with the base temporarily removed) so you can see how to connect the sensor and motor.

Controlling the Cards

Now that the robot is mostly assembled, you can add some additional protection to make sure the robot only deals one card at a time. As is, the cards will sometimes come out in large clusters, or the entire deck can sometimes slide off all at once. Gravity helps by leaning the deck toward the back of the robot, but you could improve things by adding a weight to keep the cards down or some sort of gate to keep large stacks of cards from being pulled all at once.

I tried several variations for this engineering task, using experimentation to determine what is most effective. Ultimately, using a caster wheel worked very well (see Figure 10.17). It provided enough weight to keep even, downward pressure on the cards, but it didn't prevent cards from being dealt.

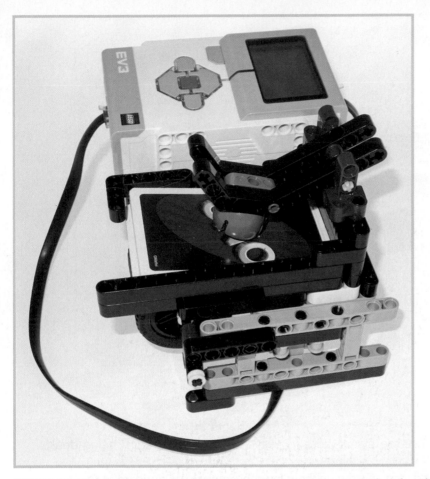

FIGURE 10.17 A caster wheel on an arm provides a solution for the card deck.

Unfortunately, a caster wheel is not a standard part of your EV3 Home Edition. You can purchase one for around $35, but if that seems wasteful for a single project there are other options.

You could try adding a sword to try to keep the whole deck from sliding off the wheel at once (see Figure 10.18). Although it did prevent the problem of dealing multiple cards at once, I found that it tended to add too much weight to the end of the deck in spite of everything being tilted backwards. When there were only one or two cards left in the deck, the cards would tilt too far away from the sensor to be read.

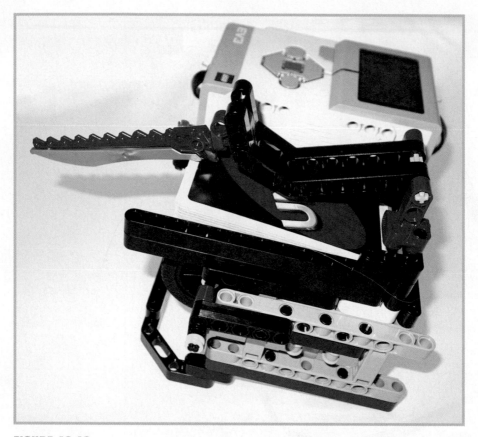

FIGURE 10.18 Here the sword acts to prevent extra cards from being dealt.

The solution I found most effective (without investing in parts not standard to the Home Edition) used just two double-bend beams to provide a little light pressure (see Figure 10.19).

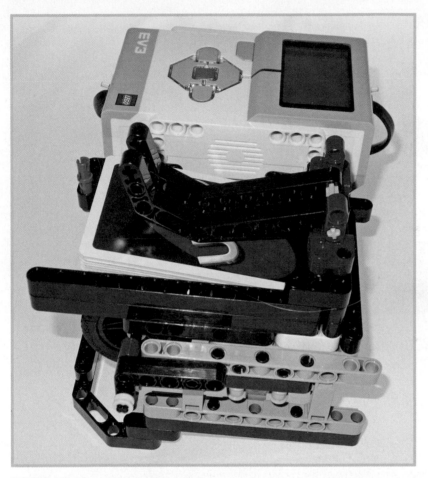

FIGURE 10.19 The double-bend beams add a bit of pressure to keep the deck in place.

In addition to having pressure from the top, you can add a 90-degree beam with a long blue peg to wrap in front of the deck to restrict the flow of cards so the whole deck can't slide off at once (see Figure 10.20).

Adjust
height

FIGURE 10.20 Adjust the height of the blue peg carefully, so only one card can fit underneath it.

You can see the entire card-holding assembly, by itself, in Figure 10.21.

FIGURE 10.21 You can lift the double-beam arms all the way up to place a deck of cards.

To make the top part of the card holder, you need to make two hinges:

1. Place two red cross blocks (one split cross block and one regular) together and secure them with a red 3M axle, as shown in Figure 10.22.

FIGURE 10.22   Use the black peg to attach the two sides of the assembly to the supporting beams. Repeat twice. There are two sides to the hinge.

2. Use a black peg to separate two double-bend beams. Place an axle through the two double-bend beams as shown in Figure 10.23.

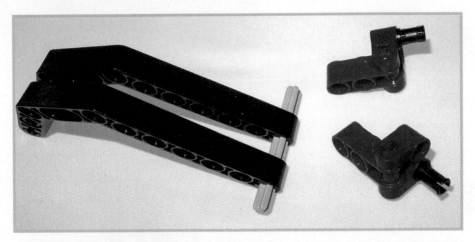

FIGURE 10.23   The double-bend beams and axle are ready for assembly.

3. Place the axle through the lower holes of the cross beams, as shown in Figure 10.24.

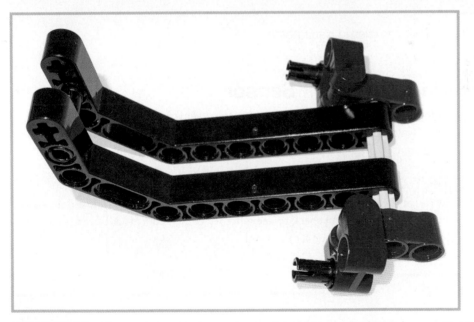

FIGURE 10.24   Place the axle through the lower holes of the beams. Putting them in the higher beam places them at the wrong angle.

4. Use an extra right-angle beam to complete the card holder, as shown in Figure 10.25.

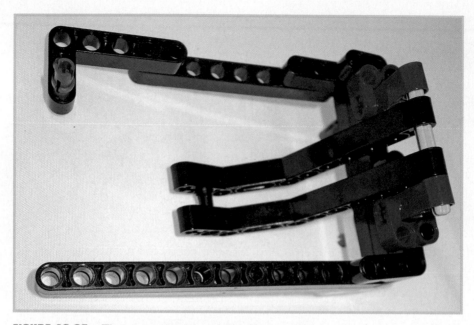

FIGURE 10.25   The completed card holder.

Calibrating the Color Sensor

I would love to tell you that you could calibrate the color sensor when it senses particular colors in the same way that you can when it searches for shades of gray with the reflected light mode. Unfortunately, you can't. The color sensor on the EV3 distinguishes among eight different colors, if you consider "no color" to be one of the eight. However, it is very precise in the color range that makes up each of the colors it senses, and you can't reprogram that range through the EV3 Home Edition software. Calibrating the color sensor, in this case, means you just have to check to see whether or not your deck of cards fits within the color range that the sensor is detecting.

> **NOTE**
>
> Advanced users can download new firmware and a new custom color sensor block thanks to the MindCuber instructions available at http://www.mindcuber.com/. The MindCuber is a Rubik's cube–solving robot that you can make from the EV3 Home Edition, and the programmed blocks go beyond the standard capabilities of the EV3 Home Edition software.

To calibrate your sensor:

1. Make sure your sensor is connected to port 3.

2. Power on the Intelligent Brick.

3. Press the right navigation button on the Intelligent Brick until you navigate to the Port View tab.

4. Press the center button to select Port View.

5. Press the right navigation button on the Intelligent Brick twice to navigate to port 3. It should say "COL- REFLECT" indicating that the sensor is currently in reflective mode.

6. Press the center navigation button on the Intelligent Brick. This lets you switch modes.

7. Press the down navigation button twice until COLOR is highlighted.

8. Press the center button to choose color mode.

9. The sensor should display a "0" indicating that no color is present.

10. Place a card in front of the sensor. The number should change to indicate the color, as shown in Figure 10.26

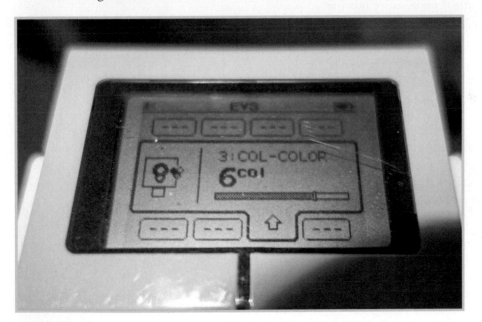

FIGURE 10.26   The color sensor is indicating 6, or white.

The sensor numbers correspond to the following colors:

0 = No color or transparent

1 = Black

2 = Blue

3 = Green

4= Yellow

5 = Red

6 = White

7 = Brown

Not all colors are on this list, so some colors will show up as a different color. Orange may be yellow, or it may be red, for example. Take note of any differences in your deck. If the differences are unique (purple shows up as red, but you don't have a red) you can work around them. If the sensor cannot tell the difference between two colors (both red and blue show up as blue) you will have to pull a few cards from your deck to make sure every card is detected at a unique sensor number.

You should not have any cards detect as white. I used white as a misalignment detector. If the sensor detects the white edge instead of the colorful section, then a card is not properly seated.

Creating the Program

Before you get started creating the program, quickly review the flowchart in Figure 10.1. The first step after the Start block is to detect the color. You could use the Color Sensor block, but this is a test you need to repeat over and over again. The flowchart loops around to keep detecting colors, so maybe the first block after the Start block should be a Loop block. Go ahead and drag that block into place (see Figure 10.27).

If you're not yet comfortable working with the programming interface, make sure to review Chapter 7, "Make Your First EV3 Program."

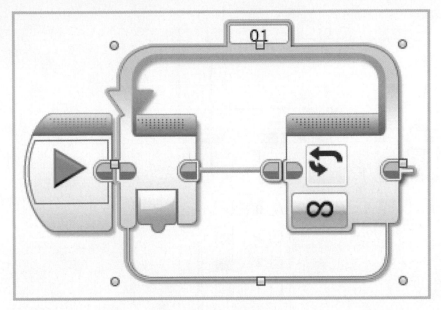

FIGURE 10.27 The humble Loop block beginning.

Because you know that the loop will eventually end when the deck of cards has all been dealt, change the mode to Color Sensor, Compare, Color. (Remember, you can change the mode by clicking on the mode area of the block.) Select 0 or "no color" as the color (see Figure 10.28). Be careful to only select 0. (You are able to select multiple colors, but don't do it for this project.) Also, double-check that the sensor port is set to 3.

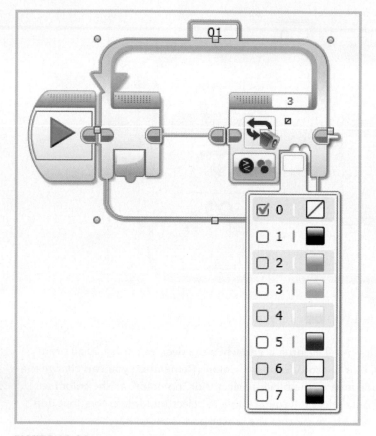

FIGURE 10.28 Be sure to select only 0.

Now that this step is complete, your loop will end when no color is detected, but it will keep looping as long as there is a color to detect.

Detecting the Color

Next on the agenda is detecting a color. Again, you could just use the Color Sensor block for this, but you need to take an action depending on which color is detected. You could use a series of Switch blocks. Actually, you can simplify this step by using one of the Switch block's secret powers. First, drag a Switch block into the sequence, and change the mode to Color Sensor, Measure, Color (see Figure 10.29).

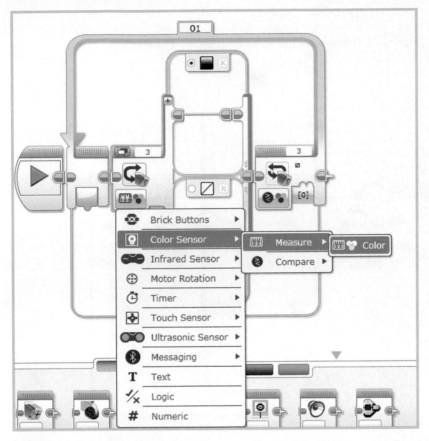

FIGURE 10.29 Add a Switch block to your sequence.

By default, the two cases are black or no color. Because you already have an event associated with no color, click on the color square and change the case to white. Remember how I mentioned the super power? Up until now, you've only used the Switch block for binary situations. Something was either true or false. You can actually add more cases than that.

Click on the subtle plus button on the upper-left side of the Switch block (see Figure 10.30).

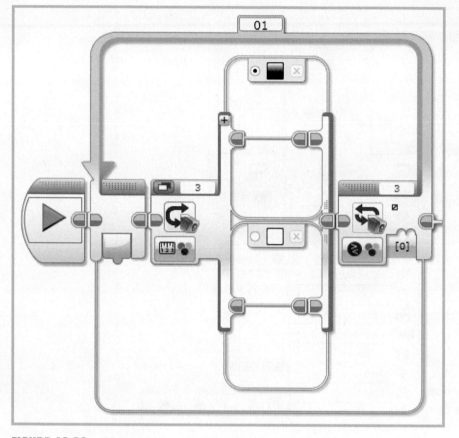

FIGURE 10.30 Click on the plus to add another case.

You can change this case to "blue" or whichever card you want to handle next. You can keep going and add the rest of your card colors as cases.

You may notice even with just one extra case that the loop is starting to get very long and a little unwieldy. If you want, you can clean it up a little by using Tabbed view. This is another easy-to-miss button on the Switch block (see Figure 10.31).

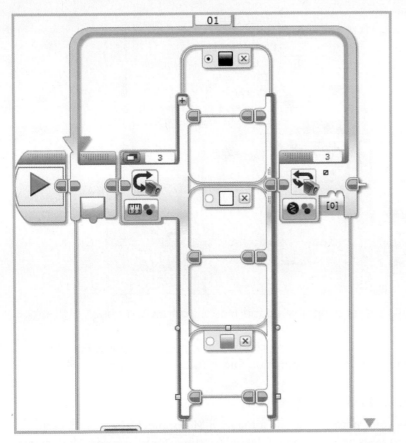

FIGURE 10.31   Click on the Tabbed View button to tame this long Switch block.

In Tabbed view, all the tabs take up the same amount of space as one case did previously, and you can just click on the left and right arrows next to the tabs to switch between tabs. Click on a tab to select it. You can always click the Tabbed View button again to switch back if the view is confusing.

Playing the Sound

Now that you've wrangled your switch cases, refer to the flowchart in Figure 10.1 to see what should happen next. You can see that each color should cause the robot to play a sound that says the name of the color and then advance the roller to deal the card. The only exception is white, which reverses the roller (to realign the card).

To set up the sounds to play, drag a Sound block into the appropriate switch case. Choose Play File as the mode. For the filename, choose LEGO Sound Files, Colors, and then find the color that matches the switch case. In Figure 10.32, the switch case is black, so the Black sound file is selected.

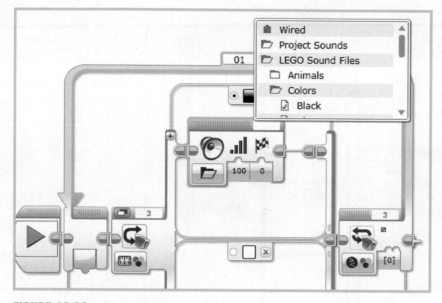

FIGURE 10.32   Choose the appropriate sound file for the switch case.

Drag Sound blocks into each switch case, and choose the appropriate sound. For white choose the Alarm sound file from the Information folder inside the LEGO sound files.

Adding the Motor Block

Now you need to make your motor work. Add a Medium Motor block after your Switch block. You'll need to test your program to find the right settings for your cards. The robot should deal one card and leave the rest of the stack properly aligned. On my robot, the settings for this task worked out to be 61 power and on for 0.2 seconds, as shown in Figure 10.33.

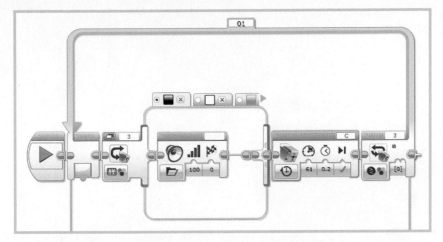

FIGURE 10.33 The medium motor has just enough oomph to deal a card.

To realign misaligned cards, add a Medium Motor block to the white case that turns for .2 seconds at –61 power. That means it will turn in the opposite direction.

Test your robot. Load your deck of cards and play the program. Ideally, it will go through the deck of cards, identify the color of the bottom card before dealing it out (using the sound you selected for that color), and then move to next card in the deck. The program will then stop as soon as all the cards have been dealt. If the cards fail to align properly with the sensor, the wheel should back up the bottom card of the deck and attempt to re-identify it until it is able to do so successfully. You may run into a few skipped cards along the way. Experiment to see whether you can come up with solutions that, as much as possible, keep the cards from being skipped.

Summary

Coming up with an idea for a robot is only part of the challenge of building them. As you learned in this chapter, sometimes it takes a lot of trial and error to find an engineering solution to a task you want your bot to accomplish. In this chapter you built and programmed a bot to deal cards from the bottom of a deck and identify the color of the cards as it deals them. In the next chapter you'll add even more complexity and learn to program using more than one robot at the same time.

11

Daisy-Chaining Projects

You've learned a lot about how to use EV3 sensors and motors to create robotics projects, but part of the real power of the EV3 is that you can network up to four Intelligent Bricks together to create even more robotics possibilities.

For the projects in this chapter, I use a LEGO Education EV3 set and an EV3 Home Edition. However, you can combine two Home Editions or other sets. As long as you have the Bricks, you have what you need and the concepts will be the same, even if the parts are not.

EV3 Intelligent Bricks can work together by either being physically connected to one another through the USB cable (daisy chaining) or by sending messages to each other over a Bluetooth network. When robots are daisy chained together, one Intelligent Brick acts as the master controller for the sensor and motor ports of the other Intelligent Bricks. Make just one program, and all the EV3s will follow. When robots communicate over Bluetooth, each acts independently with individual programs but can be programed to change behavior or react to communications from the other robots.

The Daisy-Chain Test

The daisy-chain test is just a quick experiment to demonstrate what happens when you daisy chain two EV3s together.

1. Pull out two EV3 Intelligent Bricks and connect them together using the USB included in the EV3 kit. The USB cord has both a small and large end. Plug the large end into the side of Intelligent Brick 1 and the small end into the top of Brick 2.

2. Connect a touch sensor to Brick 2 in sensor port 1.

3. Connect a medium motor to a wheel using an axle.

4. Connect the medium motor and wheel to Brick 1 in motor port A.

Your test robot should look like Figure 11.1.

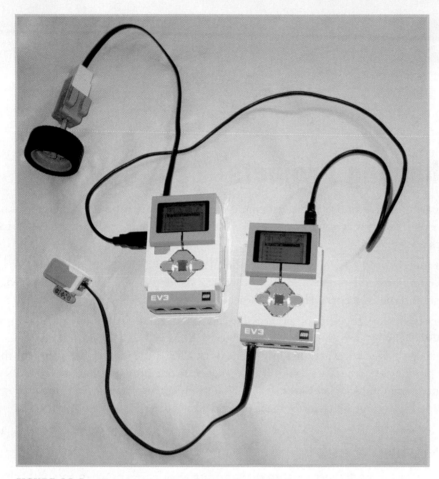

FIGURE 11.1   The test robot is ready to go.

For this program, you're not going to use the Bluetooth connection between your computer and robot. That's because, unfortunately, getting Bluetooth to work when your Intelligent Bricks are daisy chained is difficult. Instead, you can either download the program to an SD card or connect the second USB port to your computer and the Intelligent Brick.

Numbering Your Bricks

Remember how the Intelligent Brick with the medium motor is Brick 1? Just to avoid confusion, you may want to physically label the brick. A piece of tape or a sticky note will work just fine. In Figure 11.2, I used a piece of low-tack painter's tape.

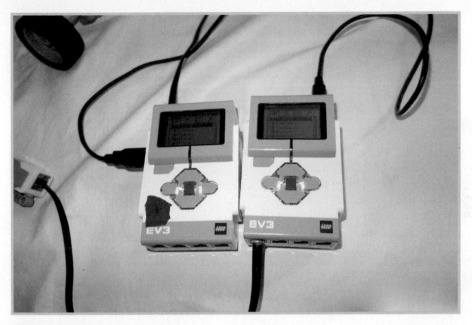

FIGURE 11.2 Using painter's tape or a sticky note means the marking is easy to remove later.

Keeping track of the Brick numbers is important as you write your test program. This program is simple, but it requires a few steps to prepare your daisy chain.

1. Launch a new project in the EV3 Home Edition software.

2. Click on Project Properties in the upper-right corner of the screen (see Figure 11.3).

testDaisy.ev3 X +

Program X +

FIGURE 11.3 The Project Properties tab looks like a wrench.

3. Select the Daisy-Chain Mode checkbox to enable the mode (see Figure 11.4).

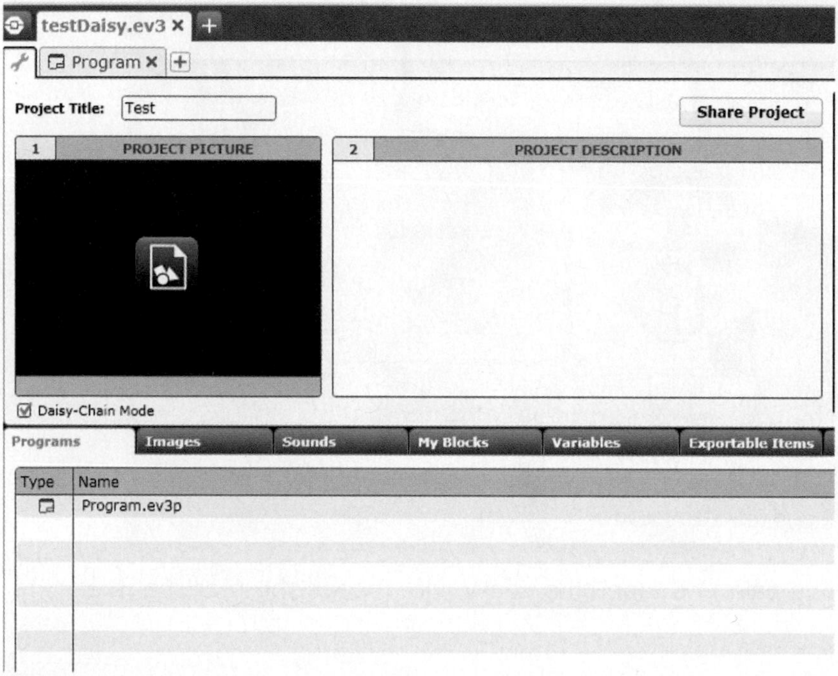

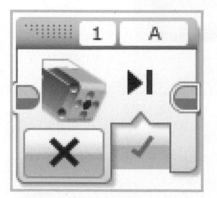

FIGURE 11.4 Daisy-Chain mode enables you to program for two or more Bricks connected together.

4. Click the Programs tab to return to the programming canvas.

You'll notice that many programming blocks now have more than one option along the top. Drag a Medium Motor block onto the canvas, for example, and you will see both a port and a layer choice (see Figure 11.5).

FIGURE 11.5 The left side designates layer, and the right side designates port.

Layers are the same as the brick number. If you click on the Layer Number button as shown in Figure 11.6, you'll see that you have four numbered choices or a "data wire" choice that can allow the layer determination to be made by data input, such as a variable. For example, you may want to randomly choose which layer will act, so you'd use a Random block to generate a number. You could also have actions determined by information gathered by a sensor.

FIGURE 11.6 The Layer number specifies which programming block controls the Intelligent Brick.

Programming the Test

Now, knowing how a daisy chain program works, you can complete a simple program with a Loop, a Switch, and two Motor blocks:

1. Drag a Loop block onto the canvas next to the Start block.

2. Drag a Switch block directly inside the loop.

3. Set the Switch block to Layer number 2.

4. Set the Port number to 1.

5. Set the switch to Touch Sensor, Compare, State.

6. Set the Mode to 1 (this is the button being pressed).

7. Inside the True value of the switch, insert a Medium Motor block.

8. Set the Mode to On, the Layer number to 1, and the port to A.

9. Copy and paste this block into the False value of the switch, but set the Mode to Off.

Your program should look like Figure 11.7.

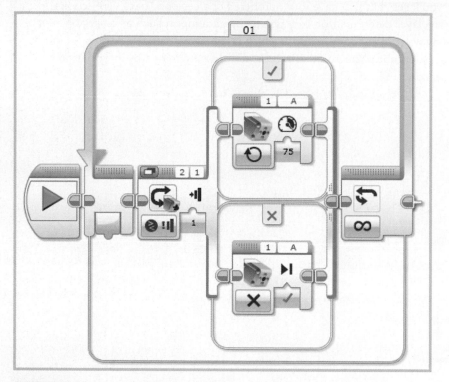

FIGURE 11.7   A simple test program for your Intelligent Bricks.

The goal is that when you press the button on the touch sensor, the motor will run. It's attached to a wheel to make the motion more obvious. The significant part of this test is that the button is on one Intelligent Brick, and the motor is on the other. Your two Bricks must communicate or the program will not work.

> **NOTE**
>
> A few blocks will not allow you to change layer numbers. The screen display, Brick buttons, brick status lights, and audio blocks are strictly controlled by the first Intelligent Brick in the chain (Layer 1). Calculations, variables, timers, and similar blocks are also controlled by Layer 1. If a block doesn't have the option to assign it to a different layer, it only works on Brick 1.

Now it is time to test the program. Hook the second USB from your computer to the port on Brick 1 and run the program. When you press the touch sensor button, the motor should turn and spin the wheel. If it does not work, try reversing your layer numbers. Make sure any physical labels you have on your bricks retain the original label.

With the touch sensor on one Intelligent Brick and the motor on the other, a spinning wheel clearly indicates the two Bricks are working together.

Building a Daisy-Chained Robot Car

Let's take this test a little further and make a basic two-wheeled car. In this version, the wheels must work together to make the car drive around in a pre-set path. Figure 11.8 shows the flowchart for this project.

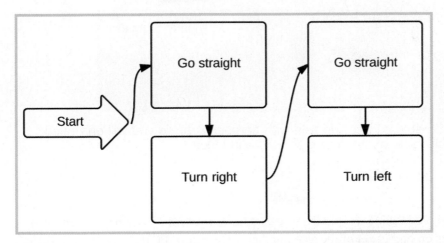

FIGURE 11.8   Your car will drive in something like a figure 8 shape.

Now let's build this car. This project uses parts from the LEGO Education EV3, the LEGO Education core expansion, and the Home Edition EV3 sets. You'll make a car with extra-large wheels and use the castor wheel to stabilize it in the middle.

Assembling the Wheels

The first task in building this bot is assembling the wheels. Just follow these steps:

1. Insert a size 8 axle with an end stop into the large motor, as shown in Figure 11.9.

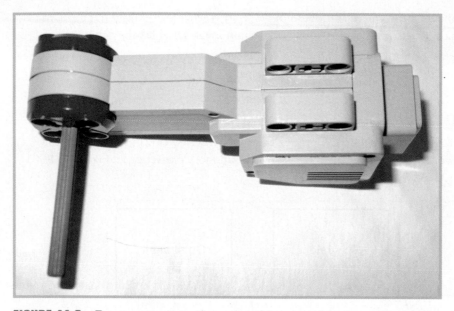

FIGURE 11.9 Be sure to use the axle with an end stop, so you will only need bushings on one side of the motor to hold it in place.

2. Add a red bushing to the axle.

3. Add a large wheel with tread onto the axle.

4. Add another red bushing to hold the wheel in place. Your wheel assembly should have red bushings on both sides of the wheel and look like Figure 11.10.

FIGURE 11.10 This wheel assembly accommodates your set's extra-large wheels.

5. Make another wheel assembly using the same steps, but this time insert the axle in the opposite side of the large motor servo, so you end up with two mirror-image wheels.

6. Gather two red cross blocks and a size 3 axle, as shown in Figure 11.11.

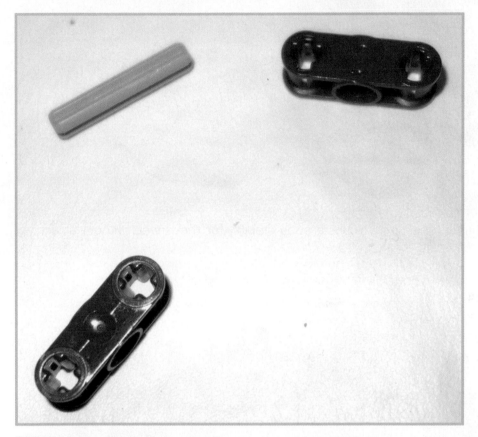

FIGURE 11.11   Be sure to actually make two sets of these, to go on each wheel assembly.

7. Place the cross blocks on either side of the axle hole on the top of the large motor servo and insert the 3M axle to keep them in place, as shown in Figure 11.12.

FIGURE 11.12 The cross blocks provide stability for the wheels without attaching them to anything.

8. Add four black pegs to a frame block, as shown in Figure 11.13.

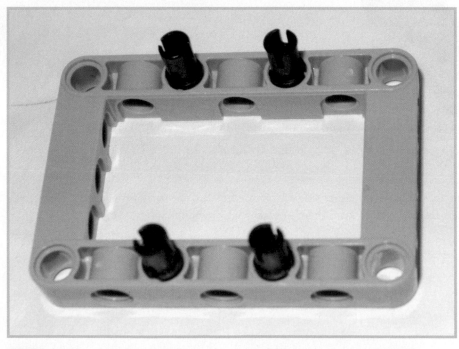

FIGURE 11.13 Add four pegs to the frame block.

9. Attach the frame to the large motor servo around the port connection, as shown in Figure 11.14.

FIGURE 11.14   Attach the beam frame to the large motor servo.

10. The entire wheel assembly should look like Figure 11.15. Repeat the preceding steps so you have two wheel assemblies.

FIGURE 11.15   Mirror this wheel assembly to make the second.

11. Plug the USB cord in the side of Intelligent Brick 1 and in the top of Intelligent Brick 2.

12. Pin both Intelligent Bricks together on the side of Brick 1 that does not have a USB cord sticking out. This step is a good example of learning by trial and error when designing a robot. Originally, as shown in Figure 11.16, I used black pegs to pin the Bricks together, but it turned out that this was actually too close together and the beams misaligned in other steps. So use long blue pegs for this step, and not the black pegs.

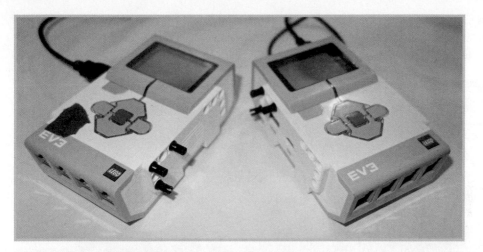

FIGURE 11.16   Both sides will easily fit together, but use long blue pegs instead of black pegs.

13. Now that you have the two Intelligent Bricks together and two wheel assemblies, use black pegs to pin the wheel assemblies to the bottom of the bot, as shown in Figure 11.17.

FIGURE 11.17   Notice how the red cross beams are actually supporting the rear part of the wheel instead of pinning any part of the robot together.

14. Put a size 15 beam across the beam frames to stabilize the robot. If, as in Figure 11.18, the beam does not center equally on both sides, you need to make sure you used the long blue pegs to attach the Intelligent Bricks together and leave a gap of exactly 1M.

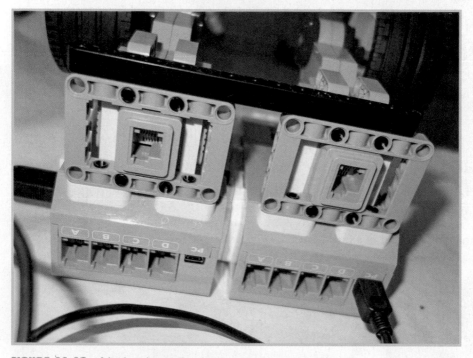

FIGURE 11.18   Notice how the beam does not fit evenly.

15. Attach the caster wheel assembly to a gray cross block and, and then add three black pins, as shown in Figure 11.19.

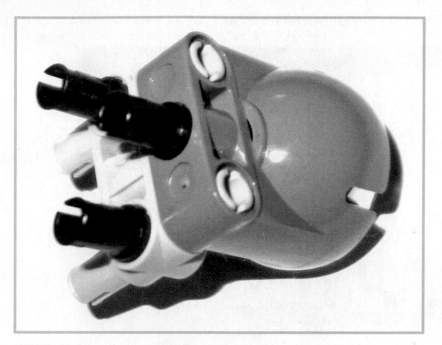

FIGURE 11.19   Insert three pegs as shown.

16. Add a size 3 beam to the cross-beam pegs, as shown in Figure 11.20.

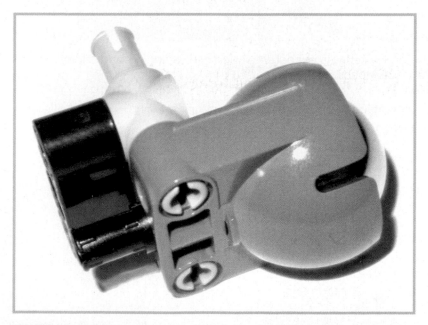

FIGURE 11.20   The extra beam is here for stability.

17. Pin the caster wheel to the middle of the size 15 beam, as shown in Figure 11.21.

FIGURE 11.21   The beam keeps the caster wheel from spinning and allows you to add components to the robot later.

18. Disconnect the USB cable and then wind the cable around the center of the Intelligent Bricks and reconnect, as shown in Figure 11.22.

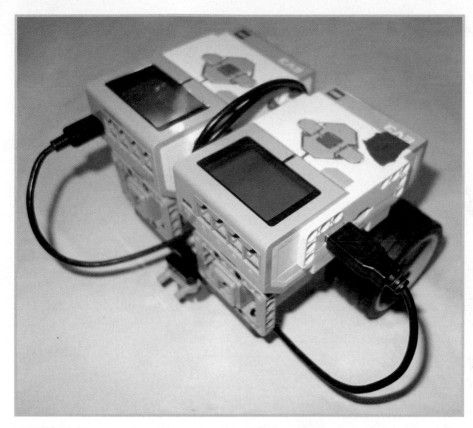

FIGURE 11.22 Winding the cable shortens your cable length.

19. Use short cables and connect each large motor servo to the D port of each Intelligent Brick.

Now that you've assembled the double robot car, it's time to program it to move.

Programming the Bot

Each motor is connected to the D port of a different Intelligent Brick, so you can't use the normal Move Steering or Move Tank blocks to control the motion. Instead, you have to use a series of individual Large Motor blocks to make this robot move.

Refer to the flowchart shown previously in Figure 11.8. Your first task is to make the robot move straight.

1. Drag two Large Motor blocks onto the canvas, and set them both to Mode: Rotation, Power: 75, Number of Rotations: 5, and Coast at end.

2. Set one block to the Layer 1 D port and the other to the Layer 2 D port. Your end result should look like Figure 11.23.

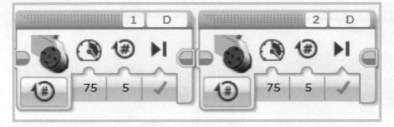

FIGURE 11.23 The two blocks control two different wheels individually.

3. Now, because you want spin, drag two more Large Motor blocks onto the sequence. Set Layer 1 to Power at 50, and set Layer 2 to Power at –50. Set the Rotations to 1 (see Figure 11.24). This makes a robot that pivots rather than traveling in a looser circle.

TIP

If you want a wider circle, one wheel needs to be slower (use less power) than the other wheel rather than moving in the opposite direction. Experiment to see what speed combination makes the optimal circle for your space.

FIGURE 11.24 One wheel moves forward while the other moves backward.

4. Copy and paste the entire sequence. Inverse the layer order for the second turn. The sequence should now look similar to Figure 11.25.

FIGURE 11.25 The entire sequence to move your car robot.

If you look at the program file, you might think it results in the bot's movement resembling a figure 8. See what kind of result you get by downloading and running the program on your EV3.

What happened? Chances are that your robot moved, but not in any way that resembled a figure 8. Mine simply spun on one wheel. Why is that? Well, the sequence executes in order, but that means only one wheel moves at a time. If the point is to get both wheels moving at the same time, then the current program won't work.

The easiest way to accomplish getting both wheels moving at the same time is to make two simultaneously executed sequences:

1. Drag a new Start block onto the canvas.

2. Keeping the blocks in the same order, drag all the Layer 2 blocks into the second sequence.

You should have something resembling Figure 11.26.

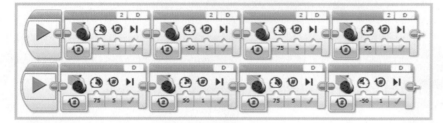

FIGURE 11.26   The simultaneous sequences make the wheels work as one.

Now try downloading and executing your program. It should now work together and make a figure 8. The lesson is that two wheels on two bricks can move as one, but only if they are running at the same time rather than sequentially.

> **NOTE**
>
> I intentionally had you make this robot complicated just to show that two wheels powered by different Intelligent Bricks could work together. In most cases, you would just want to connect both motors to the same Intelligent Brick so you can use the Move Steering and Move Tank blocks.

Adding a Remote Control

Let's take our daisy chain car a bit further and add remote control instead of a pre-set path:

1. Use a 3M axle to connect two red cross-axle beams to the base of the infrared sensor from the EV3 Home Edition.

2. Connect the cross-axle beams to a size 9M beam using black pegs.

3. Put two black pegs in the back of the 9M beam. The assembly should look like Figure 11.27.

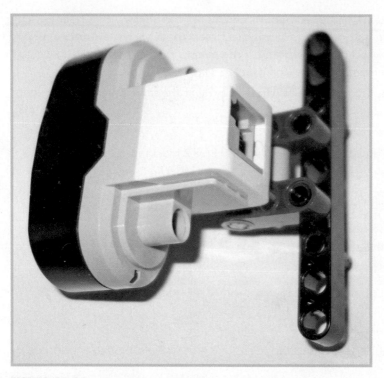

FIGURE 11.27   The infrared sensor assembly.

4. Connect the assembly to the body of the robot along the two frame beams over the two wheel assemblies, as shown in Figure 11.28 with the infrared sensor "eyes" pointing up. (In Figure 11.27, the eyes are facing away from the camera.)

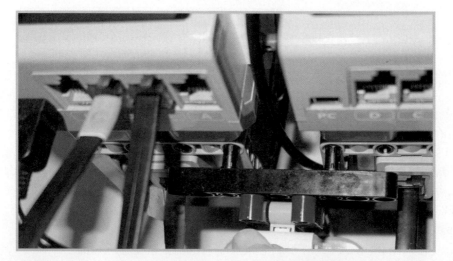

FIGURE 11.28   Connect the infrared sensor assembly to the robot.

5. Previously, you had the large servo motors connected to two different Intelligent Bricks. This was for the sake of the previous exercise. Controlling the steering from a single bot is much easier. So move the cables from the D slot in both Bricks and put them instead in ports B and C on Brick 2. This is also shown in Figure 11.28.

6. Connect the infrared sensor to port 1 of Brick 2.

Now that the sensor is connected to the robot, you can make a program that uses the remote to control the movement of the robot. You made a similar program in Chapter 9, "Engineering the Floor-Cleaning Robot," using a series of embedded Switch blocks (see Figure 11.29).

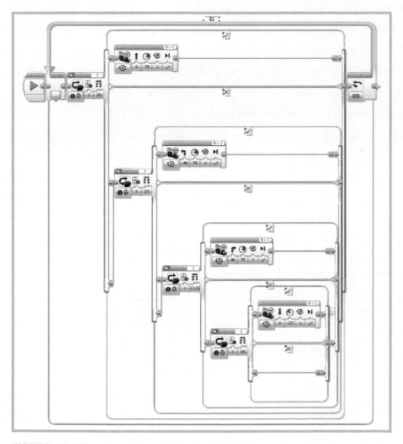

FIGURE 11.29 A series of Switch blocks makes for a very large and difficult-to-interpret program.

Let's use Switch blocks, but improve the efficiency of the program. As you learned in Chapter 10, "The Color Magic Card Trick," Switch blocks can have more than just one condition to evaluate.

NOTE

In traditional programming, a Switch block behaves like a conditional phrase, which is usually stated as an `If Then` expression, such as:

`If` (Something that can be evaluated as true or false is true.)

`Then` (Take an action.)

`Else` (If the statement is false take a different action.)

A lot of programming languages allow you add more conditions to the statement, just like the Switch block, so it becomes:

`If` (Something that can be evaluated as true or false.)

(Take an action if statement is true.)

`Else if` (A separate statement that can be evaluated as true or false.)

(Take action if statement is true.)

`Else` (If none of the statements are true take this action.)

EV3 visual programming prepares you to use this sort of programming logic when you learn to use other languages.

Using what we know of EV3 programming, let's simplify this program. Figure 11.30 shows a simple flowchart of what we want the program to do:

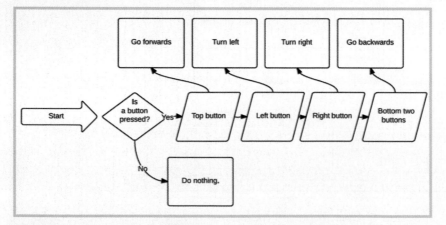

FIGURE 11.30 Your flowchart includes the concept that you don't have to evaluate each condition separately.

Now to get started with the program:

1. Start a new program in the EV3 Home Edition software.

2. Go into the Project Properties and select the check box to make this a daisy-chaining program.

3. Drag a Switch block onto the sequence.

4. Change the Mode to Infrared Sensor, Compare, Remote.

5. Your sensor is on Brick 2, so change the Layer to 2.

6. Change the Port to 1.

Now you should be able to add extra cases, and...woah, wait. As Figure 11.31 shows, you can't add cases to the Switch block in this mode.

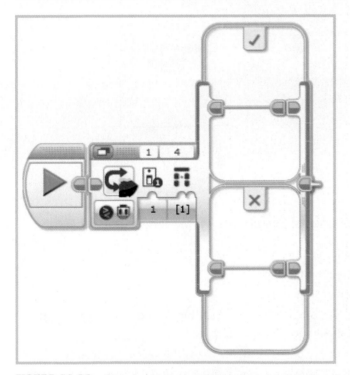

FIGURE 11.31   The button to add cases is simply missing here.

This isn't because you're working on a daisy-chained project, but rather a limitation of this particular mode on the Switch block. Don't worry. There's a way to work around it:

1. Drag the Switch block off of the sequence, but keep it on your programming canvas. This will serve as a reference as you build out the rest of your program.

2. Drag an Infrared Sensor block onto your sequence.

3. Change the Mode to Measure, Remote.

4. Check to make sure the channel setting matches your remote (this is generally only an issue if you have multiple people trying to use infrared remotes in the same area, which is why there are four channels).

5. Set your Layer number to 2 and your Port to 1. The sequence should look like Figure 11.32 at this point.

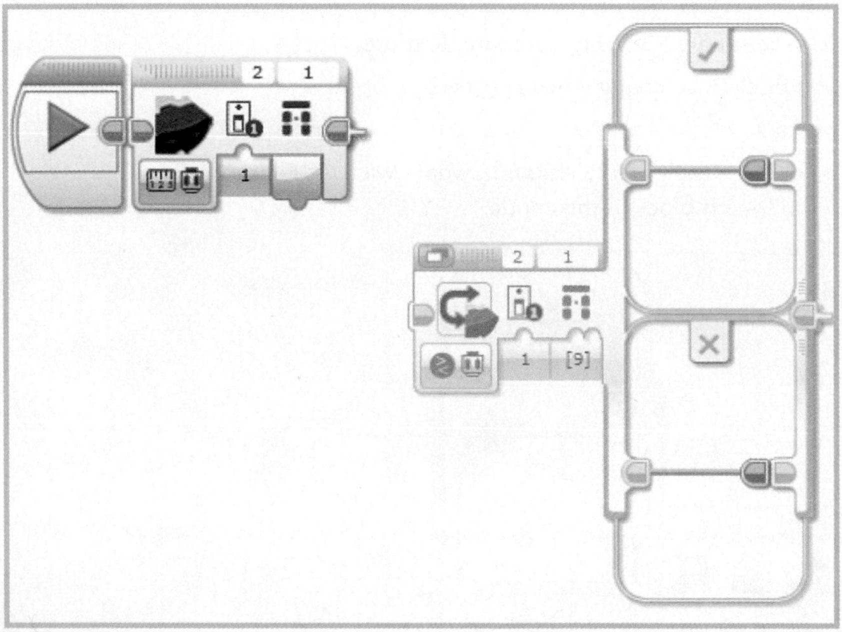

FIGURE 11.32 Notice how the original Switch block is no longer part of the sequence.

6. Drag a new Switch block onto the sequence.

7. This time, set the Switch block's Mode to Numeric.

8. Drag a data wire from the output of the Infrared Sensor block to the data input of the Switch block. Your sequence should now look like Figure 11.33.

When the infrared sensor detects a button push, it assigns the specific button (or buttons) a numeric value, and that's what gets fed into the Switch block. It's essentially the same thing as using the Switch block in Infrared Sensor mode, only now you have the option of detecting more than just two conditions.

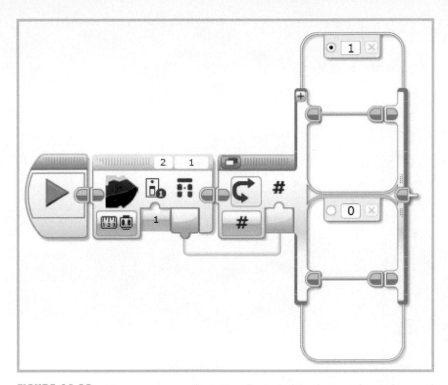

FIGURE 11.33   You can see that the Switch block now has the option to add extra conditions.

9. There are five conditions you want to detect. Add three extra conditions by clicking on the plus sign. Switching to Tabbed view might help you see things a little more clearly.

10. You held the old Switch block off to the side, so now you can use it as quick reference for the numbers. Click on the data input where you select a button to detect. Instead of selecting a button, look for the numbers assigned to it. This is your reference (see Figure 11.34).

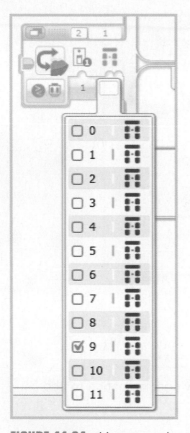

FIGURE 11.34 Here are the numbers for the button combinations.

11. Per the sensor reference, you know that the front button has the numeric value of 9, so change your first condition in the sequence Switch block from the default 1 to 9.

12. Drag a Move Steering block into the 9 value tab of your Switch block.

13. Change the Mode to Rotations, the Direction to Straight, and the number of Rotations to 1.

14. Change the Layer number to 2 and the ports to B+C. Now it's just a matter of adding the remaining states to the Switch block.

15. Copy and paste the Move Steering block. Drag it into each remaining tab.

16. Adjust the direction of the steering to correspond with the correct number.

    1: Move left –46

    3: Move Right 45

    8: Move backward (Power: 75)

    0: Nothing

17. Drag the entire sequence (other than the Start block) into an infinite Loop block. Your final program should look like Figure 11.35.

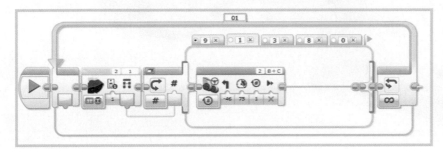

FIGURE 11.35 Your much simpler remote control program.

Adding Collision Avoidance

In Chapter 9, you also created a robot that avoided collisions by using the proximity sensor in the infrared remote. This super robot has extra ports and extra sensors, so let's improve collision avoidance by using the more accurate ultrasonic sensor instead of the infrared sensor.

First, connect the ultrasonic sensor:

1. Secure two cross-axle beams to the base of the ultrasonic sensor using a size 3 axle.

2. Secure the other side of the cross-axle beams to a beam frame using two black pegs, and then add two more pegs to the other side of the frame (see Figure 11.36).

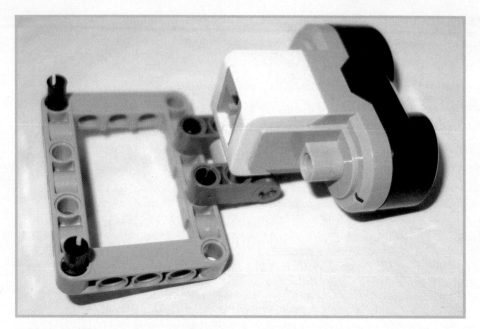

FIGURE 11.36 The ultrasonic sensor assembly.

3. Secure the ultrasonic sensor assembly to the front underside of the robot using two black pegs, as shown in Figure 11.37.

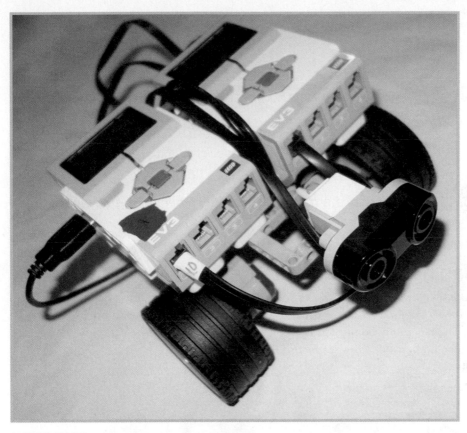

FIGURE 11.37   The collision-avoiding robot.

4. Connect the sensor to Port 1 on Intelligent Brick 1.

Now the trick is to create a program similar to what you created in Chapter 9, using the ultrasonic sensor instead of the infrared sensor.

1. Open the program you made in Chapter 9.

2. Change the Project properties to specify Daisy-Chain mode.

3. Change the first Switch block mode from Infrared Sensor to Ultrasonic Sensor, Compare, Distance, Inches.

4. Specify a Threshold value of 4.

5. Verify Layer number 1, and Port number 4.

6. Change the Layer number to 2 on all the Large Motor blocks, and verify a port value of B&C.

The end result should look like Figure 11.38.

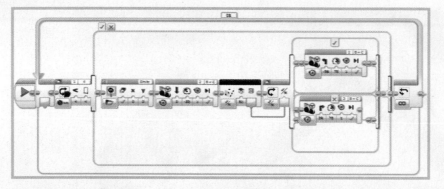

FIGURE 11.38 Because this is a modified program rather than a new program, developing it quickly is easy to do.

Messaging Between Robots

There's more than one way to get two robots to work together. You don't have to physically daisy-chain them with a USB cable. You can also have them message each other with Bluetooth. The Bluetooth Communication blocks are located in the blue Advanced tab in the EV3 Home Edition software (see Figure 11.39).

FIGURE 11.39 You use the Bluetooth Connection and Communications blocks to send messages between two separated EV3s.

> **NOTE**
>
> When you use Bluetooth communication, you aren't making multiple EV3 robots act as one robot from a single program. Instead, you must write a program for each EV3 and have it act separately. You can separate your robots physically, but only within Bluetooth communication range, which generally means within the same room and not behind walls.
>
> Renaming at least one of your EV3 robots in the Brick Information tab in your EV3 software is also important, so you can specify exactly which robot sends and receives the messages.

Adding "Magic" to the Card Trick

In Chapter 10, you made a card-dealing robot that identified the colors in a deck of Uno cards. Now you're going to modify that program to have the robot send the information to a different robot across the room. The "magic" is that an EV3 without any sensors or motors is going to know the card's color and control when the next card in the deck should be dealt. It's a mind-reading robot trick!

Prerequisites: A fully assembled robot from Chapter 10 and a second Intelligent Brick. Both EV3s should have different names, which you can set in the Brick Information area of the EV3 Home Edition software. I named mine "EV3" and "EV3 EDU."

1. Start with two copies of the color-identifying program from Chapter 10. Name one "Sending" and one "Receiving." Simply copying and pasting the entire sequence into another program tab within the same project is most efficient to keep everything contained.

2. Drag a Bluetooth Connection block onto the sequence, just after the Start block on both the sending and receiving programs (see Figure 11.40).

3. Set the Mode to On.

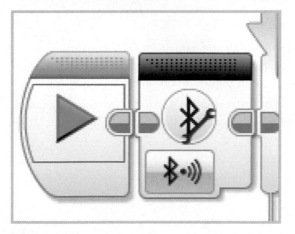

FIGURE 11.40 This turns on the Bluetooth for your EV3s.

From here you need to separately configure both the sending and receiving programs.

Configuring the Sending Program

Let's now focus on your Sending program:

1. Drag another Bluetooth Connection block just after the first one and before the main sequence Loop block. Change the Mode to Initiate and the Connect To: input to the name of the other EV3, which is EV3 EDU in this example (see Figure 11.41).

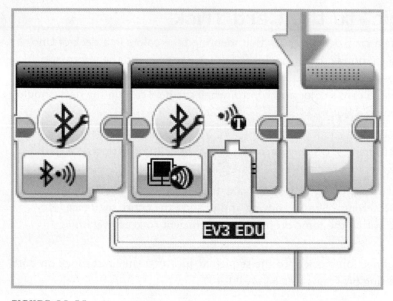

FIGURE 11.41 Type in the name of your other EV3 robot.

This establishes a connection between the two robots.

2. Previously, your program used sound to announce the name of a color before proceeding. Now you're going to have the robot silently send a text message to the other robot. Replace each Sound block in your switch sequence with a Messaging block.

3. Set the Mode to Send Text. Set the name of the receiving EV3, and set the message to match the color. For example, Figure 11.42 shows the message set to Blue to match the color choice of blue.

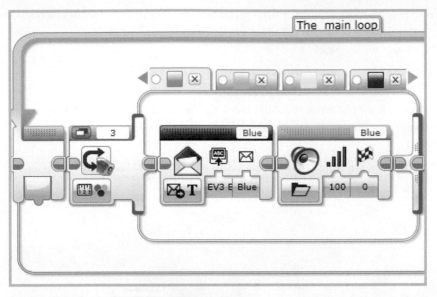

FIGURE 11.42 Also name the text message to avoid future confusion.

4. Delete the Sound block after you get your new Messaging block in place. If you copy and paste the Messaging block, you only have to make a few tweaks instead of changing all the settings each time.

5. After you've replaced all the conditional tabs in your Switch block, add a Wait block. Change the Mode to Messaging, Compare Text. Set it to equal Next (see Figure 11.43).

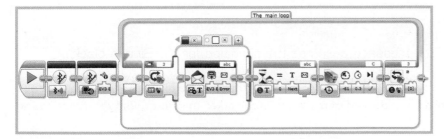

FIGURE 11.43 This sequence sends results as a Bluetooth message and waits to hear "Next" back from the other robot before proceeding.

Configuring the Receiving Program

Now let's move on to the Receiving program. You should have already added the Bluetooth Connection block next to the Start block.

1. Change your Switch block mode from Color Sensor to Text. This changes your conditional tabs to numbers (corresponding to the colors from the color sensor).

Change them to text that corresponds to the messages your Sending program will be sending. So, "Black," "Blue," "Green," "Red," "Yellow," and "Error." Make sure you match the text to the proper Audio block.

2. Add a Wait block just before your Switch block. Change the Mode to Messaging, Update, Text. This tells the robot to wait until a new text arrives.

3. Connect the data wire output from the Wait block to the input on the Switch block. This sets the text in any Bluetooth message to be the conditions for the Switch block (see Figure 11.44).

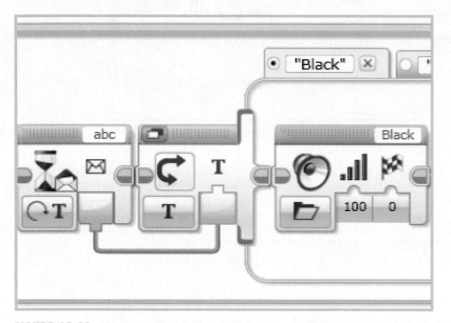

FIGURE 11.44 The message becomes the condition for the Switch block.

4. Now you can delete any of the previous blocks that occurred in the sequence between the Switch block and the end of the loop. They worked the motor, and this robot has no motor servos.

5. Add a Display block after the Switch block set to Text, Grid. Add the message "Press Center Button."

6. Add a Wait block after the Display block. Set the Mode to Brick, Buttons, Compare, and set it to detect the center button being pressed and released (2, 2).

7. Add a Messaging block by setting Mode to Send Text. Name the sending robot, and set the text of the message to Next.

8. Add a final Display block set to Reset Screen.

Your final sequence should look like Figure 11.45.

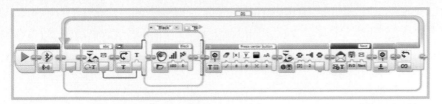

FIGURE 11.45 The final Receiving program.

Running the "Magic"

With both Sending and Receiving programs complete, load both programs onto their respective bots, and try launching.

If everything goes according to plan, the card-dealing robot should detect the color, but make no sound. Instead, it sends a Bluetooth text message to the other robot across the room, which says the name of the color and waits for you to press the center button. After you do, the card-dealing and color-detecting robot will deal the next card and attempt to identify the color.

Summary

In this chapter you explored several ways in which two or more EV3 robots can communicate with each other and run programs either in tandem or as a daisy-chained "super" EV3. You learned how to program a Switch block more efficiently to detect multiple conditions, and you learned how to send Bluetooth messages between robots. In the next chapter, you'll explore alternative programming languages and ways to expand the EV3 with third-party parts.

12

Extending Play

In this chapter, you'll explore more ways to extend EV3 play. I walk you through the steps to boot the alternative EV3 programming language leJOS, so you can program the EV3 with Java. This chapter also provides a look at several bot models created by the LEGO community, in the hope that they will inspire you. This chapter also provides tips and resources for finding and getting involved in LEGO communities and LEGO competitions, as well as some alternative sources for spare parts.

Installing leJOS

As I mentioned in Chapter 1, "What's in the Box?" the EV3 allows you to use alternative programming languages and operating systems if you so choose. As this book is being written, one of the most fully developed systems for the EV3 is leJOS, a port of the Oracle Java programming language. Java is designed as a write once, run anywhere language, because programs run within a "virtual machine" that doesn't require you to recompile programs for different platforms. leJOS acts as this virtual machine for EV3, so programs written on your desktop computer in Java will run on your EV3 with leJOS. Although leJOS is not an official Oracle product, Oracle offers links on its site to the leJOS community for users interested in EV3 programming.

> **NOTE**
>
> **Why leJOS?**
>
> leJOS is free, open source, and actively supported. It's also easy to install and uninstall, and it doesn't require root user access to the LEGO EV3 operating system. By using an alternative programming language like leJOS, you can more easily transfer your programs or components of your programs to other Java robots.

Previous versions of leJOS required users to download components from GitHub and go through a lot of complicated steps. (GitHub is a code management tool used to keep track of programs with multiple contributors.) As of this writing, you no longer have to use GitHub, but you do have to go through a few steps to create a formatted SD card. Hopefully by the time you're reading this, the process has simplified further.

Preparing Your Desktop

Before you can use leJOS, you need to prepare your computer desktop programming environment:

1. Download the latest developer version of the Java Development Kit from Oracle.

> **NOTE**
>
> The Java Development Kit (JDK) is not available from Java.com. Java.com offers the Java Runtime Environment (JRE), which is what you use to run programs written in Java on your computer. The JDK allows you to actually write programs, and it includes the JRE.

2. Download Eclipse from www.eclipse.org. This open source, integrated development environment (IDE) makes writing Java programs for your EV3 a lot easier than coding them in a simple text editing program. Technically, this step is optional, but I recommend it. Eclipse can help you find typos before they become programming errors and offers options to help you keep track of your programs as you write them.

3. Launch Eclipse. You need to set a workspace, but it doesn't really matter where it is at this point, because you're not going to write a program yet. The default is just fine.

4. Go to Help, Eclipse Marketplace, as shown in Figure 12.1.

FIGURE 12.1 You can find the EV3 plug-in in the Eclipse Marketplace.

5. Use the search filter to find leJOS, as shown in Figure 12.2. Be sure to select the leJOS EV3 beta, not the NXJ version of the software.

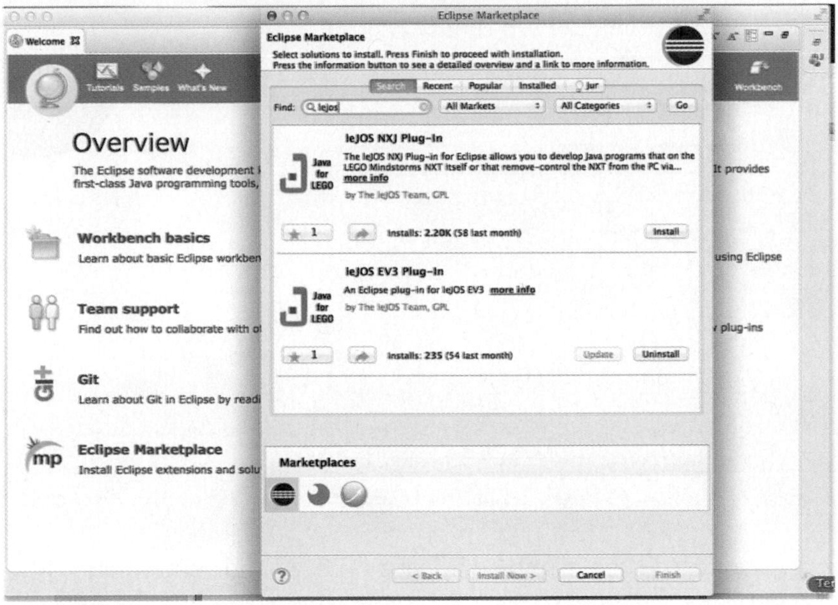

FIGURE 12.2 You could also search for "EV3" instead of "leJOS."

6. Click Install. You might see a warning that this plug-in is unsigned, but confirm that you want to install it anyway. It's safe.

7. Restart Eclipse to finish installation.

8. You'll see a dialog box asking where you want this project to be located. Set a new location for your projects if you want them to be anywhere but the default location. (Using the default location is fine, if you prefer.)

9. Check option in this dialog box to make this the new default location if you don't want to keep seeing this dialog box every time you launch Eclipse.

10. Go to Preferences.

11. Select leJOS EV3.

12. Set the EV3_HOME by browsing to the location of your plug-in installation, as shown in Figure 12.3. Click OK.

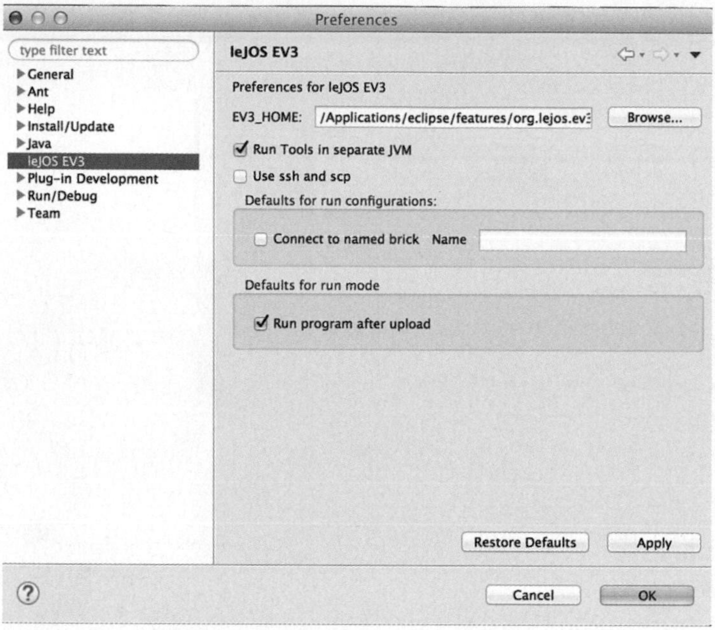

FIGURE 12.3 The location might be slightly different on a Windows machine, but the concept is the same.

Loading the SD Card

Now that you've set up your programming environment, you need to load an SD card and boot leJOS onto your EV3. You need a micro SD card, such as the one shown in Figure 12.4.

FIGURE 12.4 The micro SD card is shown next to the adapter.

Micro SD cards are used in some smartphones and tablets, and they usually have a larger SD card adapter, so you can put them in standard-sized card readers, which are common to laptops and desktop computers. You can also purchase a card reader separately.

Your micro SD card can't be bigger than 32 GB, and you need some way for your desktop computer to write to it, such as the adapter shown in Figure 12.4.

TIP

Before following these instructions, verify that the instructions have not changed by referring to the leJOS wiki at www.legos.org.

These steps work for leJOS beta version 8.1:

1. Download leJOS from http://sourceforge.net/p/lejos.
2. Download version 7 of Java SE Embedded JRE from Oracle at http://www.oracle.com/ technetwork/java/embedded/embedded-se/downloads/index.html#javase7update. You must register for an Oracle account to do this download, but registration is free. You have a lot of choices for downloads, but as of this book's publication, the file you should download is the RMv5 Linux - Headless EABI, SoftFP ABI, Little Endian.

NOTE

Yes, you're downloading another version of Java, but this version runs on embedded devices, such as compatible watches, car dashboards, or (in this case) EV3 robots. Java and the leJOS components are separate downloads for legal reasons. leJOS is not an official Oracle project, but Java is.

3. Format your micro SD card, if necessary. It should be formatted to FAT32. Most likely you won't need to do this because micro SD cards are already formatted to FAT32 by default. Your computer will prompt you to format the card if you insert an unformatted card into your card reader.
4. Open and expand your leJOS download.
5. Inside the leJOS folder, you should have a file called lejosimage.zip. Unzip that file directly into the primary folder or root directory of your micro SD card.
6. Copy the JRE.gz file (that's the one you downloaded from Oracle) onto the card.
7. Insert your prepared micro SD card into your EV3.
8. Turn your EV3 on, and the leJOS boot screen appears, as shown in Figure 12.5.

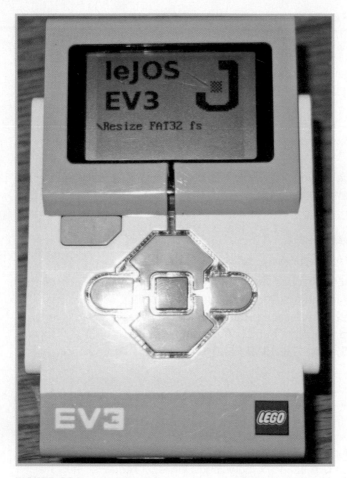

FIGURE 12.5 The leJOS boot screen lets you know you have the micro SD card installed.

The first time you boot up using leJOS, it takes close to 10 minutes for everything to be formatted. The screen should let you know it's progressing. Future boot ups are much faster.

If you were unsuccessful in creating a bootable SD card, one of two things will happen. Either nothing will happen and the EV3 will boot up and never show you a leJOS logo, or leJOS will boot up and even after 10 minutes will remain unresponsive. If that happens, give it 10 more minutes just to make sure, but don't worry too much. You have not broken your EV3.

To handle an unresponsive SD card:

1. Remove and reinstall the batteries from your EV3 to turn it off.
2. Remove the micro SD card from the side of the EV3.

3. Restart. You will now boot up to normal EV3 environment. That's one of the perks of using leJOS. No matter what happens, you aren't removing the normal EV3 operating system.

4. Reformat the micro SD card, and try again.

If you continue to have problems, verify that you do not have an SD card bigger than 32 GB, which isn't supported, and double-check that you've followed all directions for creating the micro SD card image for leJOS.

Working in LeJOS

If all goes according to plan and you boot up your EV3 using leJOS, you'll see that you now have a different menu system than the default EV3, as shown in Figure 12.6.

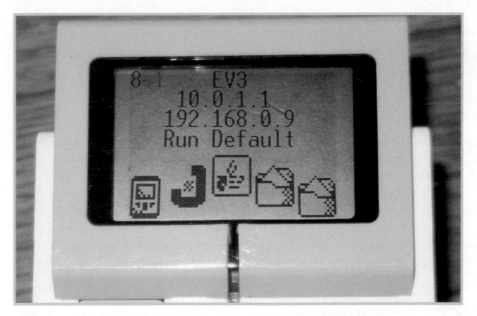

FIGURE 12.6 Navigate using the EV3 Intelligent Brick buttons.

You can navigate between items by using the EV3 Intelligent Brick buttons. Left or right buttons navigate through top-level menu items, and the center button either executes a program or opens a file list, depending on where you navigate.

An in-depth tutorial on programming in leJOS is beyond the scope of this book, but here is a general overview. Programs that run on Java have the .jar extension and can be compiled using Eclipse or another IDE. Programming Java files is done all in text, although Eclipse does provide you with code hints and debugging tools to make it an easier journey. In the EV3 Home Edition software, you see programs as interlocking blocks, but when using Java, you have to know how and when executing a command is appropriate, and you have to type it all out rather than dragging and dropping.

For example, a Java command to tell the screen display to show the classic "hello world" message would look like this:

```
TextLCD.print("HELLO WORLD");
```

It is doing the same thing as a Display block set to display the text "HELLO WORLD." leJOS is a more complex environment, but it's also more powerful.

After you have created an entire program, you need to compile it into a .jar file and write it to the micro SD card for it to work.

To take apart a program that someone else has already written, use one of the sample programs available from the leJOS Git repository. This currently includes:

- EV3BumperCar
- EV3ColorTest
- EV3GraphicsTest
- EV3SensorMonitor

You can also find information and tutorials on the sample programs from the leJOS wiki at http://sourceforge.net/p/lejos/wiki/Home/.

Community-Created Models

In Chapter 4, "Building Your First Bots," we went through the default model robots created by LEGO, but the community has always been part of the idea behind LEGO robotics. LEGO robotics enthusiasts have always informally created groups and forums to share ideas, and LEGO has created an official channel to share ideas with other members of the LEGO robotics community as well. Community shared robots can be downloaded through the Community link in the Lobby area of you EV3 desktop software. You can submit your own robots to the community through the LEGO Home Edition software by following these steps: by following these steps:

1. Open the project you want to share.
2. Go to Project Properties.
3. Add a description.
4. Add at least one project picture.
5. Optionally, you can add a video of your robot in action.
6. Click on the Share Project button.

Prior to submitting a robot, make sure you have excellent documentation on your build. Figure 12.7 shows the Share screen.

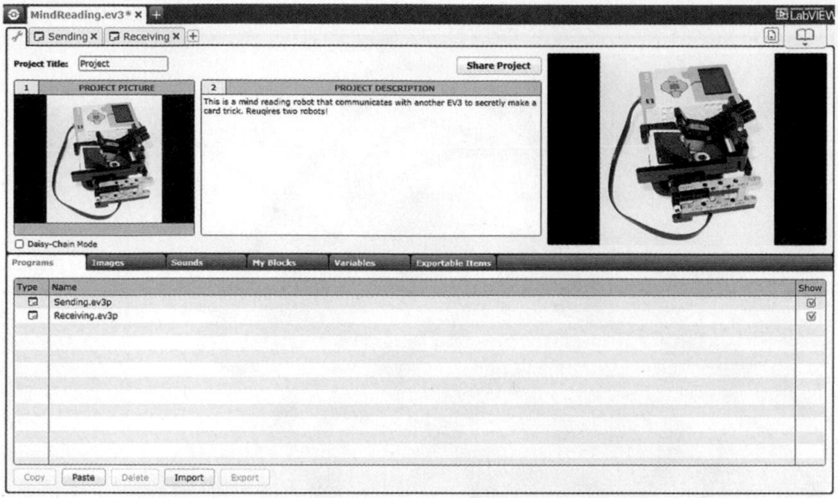

FIGURE 12.7   This project is ready to share with the community.

LEGO also seeded the community with several contributions from early testers, and those robots are also worth checking out. You can get to these contributions by going to the Home Edition software lobby area and clicking More Robots. Let's have a look at some cool examples that spawned from the Lego community.

DINOR3X

DINOR3X is a crawling dinosaur of a robot (see Figure 12.8) and is worth checking out to see how the legs are engineered to make the robot move without falling over.

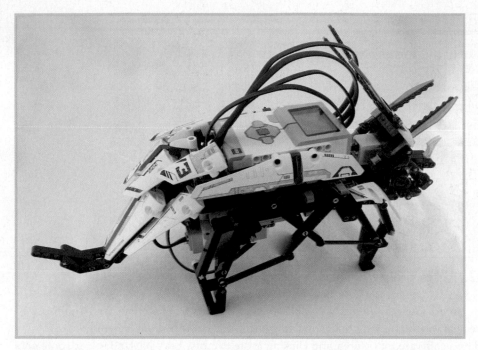

FIGURE 12.8   DINOR3X is both very stable and has fierce dinosaur looks.

EL3CTRIC GUITAR

The EL3CTRIC GUITAR is one of my favorite bonus builds. It's an electric guitar that makes different pitched sounds based on the location of the sliding wing part shown in Figure 12.9. Strum it by pressing a lever that presses down on a hidden touch sensor.

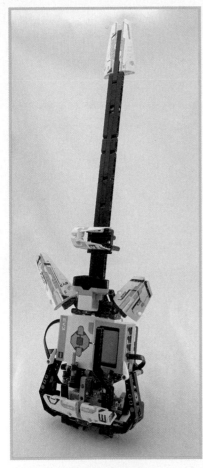

FIGURE 12.9 Notice how the gray axle near where guitar strings would normally be found is carefully positioned to press down on a hidden touch sensor.

EV3D4

EV3D4 is an R2-D2 style robot with a "space ship" remote control, as shown in Figure 12.10. If you examine the program that comes with the EV3D4, you'll notice a tabbed Switch block and a very interesting block controlling the pulsing of the buttons on the Intelligent Brick in response to remote pushes.

FIGURE 12.10   The robot gives feedback on remote signals by pulsing the button lights on the Intelligent Brick.

EV3MEG

The EV3MEG is a robot with working arms and grasping pincers (see Figure 12.11). This is another great engineering example to build if you want to figure out how to make functional robot arms on your EV3.

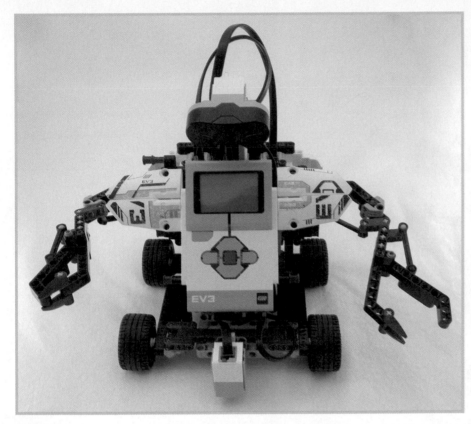

FIGURE 12.11 Four wheels and two arms is actually quite an accomplishment for the limited number of parts in the EV3 Home Edition.

MR B3AM

MR B3AM, pictured in Figure 12.12, is a robot that uses the color sensor and wheels to detect and measure the color and length of Technic beams. As a practical robot, this isn't really much help, because most people are faster counting holes and visually sorting colors than it can be. However, as an engineering project it's extremely cool.

FIGURE 12.12   Insert the beam in the gap just under the wheel.

KRAZ3

KRAZ3, shown in Figure 12.13, is another robot with a decorated infrared remote/beacon as a "friend." You can either use the programmed remote functions or set KRAZ3 to follow the remote in beacon mode. What I find most interesting about this robot is the unique solution on what to do about extra-long tank treads. Rather than making a long ellipse, the tank treads become a triangle and still retain the stability of a tank. Other builds, such as BULLDOZ3R, take a similar approach.

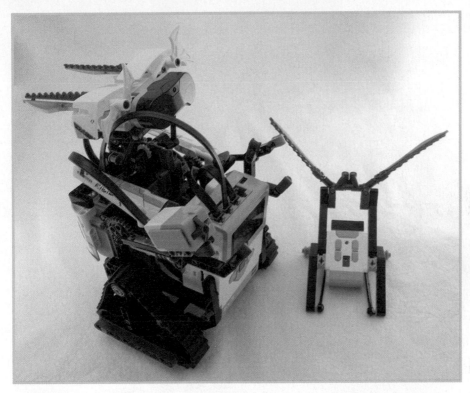

FIGURE 12.13 Notice how the tank treads take up a shorter but taller space underneath the bot without making KRAZ3 unstable.

RAC3R

RAC3R is intended as the basic building block for a racecar that can be modified with new gears to experiment with different designs (see Figure 12.14). As built, it can drive with or without remote control.

FIGURE 12.14 The RAC3R has more of a truck-like appearance, so you might want to modify it to drag other objects.

EV3GAME

EV3GAME is a shell game played with one of the red balls and three tires (see Figure 12.15). You hide the ball under a shell, use the remote to set the difficulty level, and then try to guess where the ball is hidden.

FIGURE 12.15   The EV3GAME is a giant shell game.

The shell game aspect is interesting by itself, but I encourage you to take a look at the program. By taking advantage of custom blocks, the program used for EV3GAME is super short and clean, as shown in Figure 12.16.

FIGURE 12.16   Look at this single line of code with an optional End block.

WACK3M

WACK3M, shown in Figure 12.17, is a "whack-a-mole" game that uses tires as pop-up "moles." What makes this game interesting is the use of the infrared sensor to detect hits rather than using a touch sensor.

FIGURE 12.17 WACK3M is also fun because the build instructions contain the warning not to hit anyone in the head with the mallet used for construction.

BANNER PRINT3R

BANNER PRINT3R is a robot you can use in combination with a magic marker and receipt printer paper (or similarly sized strips of other paper in a pinch) to draw banners on the paper (see Figure 12.18). That, in and of itself is pretty cool, but the significant thing about this particular model is that it served as an inspiration for 12-year-old Shubham Banerjee to solve a real-world problem. When he found out how expensive Braille printers were, he modified the BANNER PRINT3R to stab Braille texture into paper rather than write on it with a magic marker.

FIGURE 12.18 This printer served as inspiration to create a Braille printer known as the Braigo.

If you want to do something similar, you can find the instructions for building your own Braigo on MAKE: http://makezine.com/projects/braigo-a-diy-braille-printer-with-lego/.

Finding More Communities

The LEGO community doesn't exist solely on LEGO's EV3 forums. You can find sites devoted to LEGO Robotics, to finding parts, and even to specific robots. The MindCub3r, for example, is a Rubik's Cube–solving robot, and you can download the instructions from http://mindcuber.com/mindcub3r/mindcub3r.html.

MindCub3r is a port of an NXT build (The MindCuber), but the project also goes even further. An extra-intensive version with multiple EV3 parts and a mobile phone used as the "brain" was able to solve a Rubik's cube in 3.2 seconds. (The fastest human record was 5.5 seconds by Mats Valk in 2013. If you've ever seen a Rubik's cube competition, it's amazing.)

Scoring Extra LEGO Parts

As mentioned in Chapter 1, the EV3 is compatible with the LEGO Technic system. You can purchase Technic kits and use the pieces with your EV3. Increasingly, standard LEGO system kits have a few Technic parts included.

You can also buy new sensors and parts from LEGO Education, as mentioned in Chapter 3, "Comparing the EV3 and NXT," (see https://shop.education.lego.com). In addition to being able to purchase the ultrasonic sensor, castor ball set, and gyro sensor, you can purchase renewable energy kits, probes, and the Space Activity Pack.

Brick Owl (http://www.brickowl.com) is an unofficial LEGO marketplace consisting of multiple independent stores, and the focus is mainly on individual parts. If there's a part you need, there is very likely a seller and a price.

LEGO also has a parts store at http://shop.lego.com/en-US/Pick-A-Brick-ByTheme. You can filter the results to only display Technic.

eBay and Craigslist are also potential places to find extra parts. Sometimes you can buy assorted LEGO pieces by the pound from resellers who are trying to find specific mini figures or other collectibles.

Tetrix

LEGO Education and Pitsco also introduced a separate but compatible building system called Tetrix that can be controlled by an EV3. Tetrix parts are made out of metal (aircraft-grade aluminum) and enable you to make much stronger and bigger robots than what you could create with just the plastic parts in the EV3 Home Edition set.

To find more information go to http://www.tetrixrobotics.com/.

K'nex

Not all K'nex parts are compatible with the EV3, but most of them are. The electric parts are not compatible. You can use the EV3 to power what you build using your K'nex.

Erector Sets

The EV3 has some part compatibility with Erector, although probably less than 50% of the parts can be used. If you have an old Erector set lying around the house, exploring whether it will work with your build might be useful.

3D Printers

If you own a 3D printer, you can print your own pieces for the EV3. You can't print entire Intelligent Bricks or sensors, but you can print beams, axles, and other useful parts. Check out Thingverse for downloadable printer instructions here: http://www.thingiverse.com/tag:LEGO.

If you don't have or can't afford a 3D printer, you can also check to see whether your city offers a Makerspace. Makerspaces, also sometimes called "hacker spaces" are community-operated spaces that offer use of common equipment such as 3D printers in exchange for a membership fee. There are also commercial companies that will print 3D objects for you, although using a company to print it may end up costing more than buying a single brick from Brickowl.

Robotics Competitions

One fantastic way to go further with the EV3 is to compete, either alone or on a team. This gives you fresh challenges every year, and inspiration as you see how others have solved the same challenge. Generally, LEGO robotics competitions are intended for kids, but adults are always welcome as coaches and volunteers. As a volunteer or coach, you can tap into the community of adult LEGO robotics enthusiasts. (That doesn't mean you have to be a grown-up all the time. Host an informal mini-challenge with your fellow coaches.)

First Robotics LEGO League

First Robotics is probably the best-known robotics competition league. First Robotics offers LEGO robotics competitions—First LEGO League (FLL)—up to eighth grade. You can either find a group or create one by going to http://www.usfirst.org/roboticsprograms/frc.

In addition to competitions, First Robotics offers great resources for potential coaches and discussion forums for troubleshooting EV3 programming or design issues.

World Robot Olympiad

The World Robot Olympiad (WRO) is an international, school-based competition involving groups of three students competing together. You can find more information at http://www.wroboto.org/.

4-H

The 4-H organization has a renewed interest in teaching Science, Technology, Engineering, and Math (STEM) to children between the ages of 7 and 18. Many resources and categories depend on the state and county in which you reside, but 4-H has developed a robotics project and most counties offer robotics competitions as part of the county fair. In many counties and states, competition can be a team effort or that of just an individual.

To find out more about 4-H or the location of a club near you, visit them at http://www.4-h.org.

Decorating Your EV3

Nobody said that your EV3 had to look just as it did when it arrived in the box. If you want to decorate your EV3, you can use model paints, stickers, duct tape, fingernail polish, and many other tools. Spray paint is one of the fastest and most permanent ways to decorate EV3 parts.

Avoid painting sensors, wires, gears, and the Intelligent Brick. Those parts have sensitive ports and moving parts, so they can easily be jammed, fried, or otherwise damaged.

Try a patch test on a less valuable part. If you get too much paint on a piece, you run the risk that the piece will no longer fit with the other pieces.

Parts you can potentially spray paint include beams, beam frames, the white flat wing parts, the spikes, and the swords. Do not spray anything on servos, sensors, cords, tires, gears, or the Intelligent Brick. Keep those parts well away from any spray paint. When painting parts, keep the following in mind:

- Use Krylon's Fusion for Plastic spray paint or a similar brand meant specifically for plastic. That way you can avoid primer and reduce the number of layers you need.
- Spray in a well-ventilated area, such as outside.
- Spray in very thin coats.
- Avoid drips.
- Allow everything to dry between coats.
- Turn the parts after they dry, so you get all angles.

According to Krylon, the paint needs seven days to fully cure and be "chip free," so be patient when painting parts. Use them too soon and you risk wasting the effort you put into painting them in the first place. You can also use model enamel to paint LEGO parts, but again, avoid the servos, sensors, and Intelligent Brick.

Summary

This chapter explored the idea of using leJOS as an alternative to programming using the LEGO visual programming environment. You saw some of the bots created by the LEGO community and perhaps even inspired by them to come up with your own versions. You learned about multiple ways to score extra parts, including buying kits from LEGO competitors. Finally, you learned about some of the LEGO competitions and communities out there. As you learn and grow with LEGO robotics, be sure to share what you've learned with others.

Glossary

If you are new to the world of LEGO, robotics, or programming, you might see a lot of terms used in this book that may be confusing. While I've done my best to define terms when they are introduced, that won't help if you're skipping chapters and flipping to the parts of this book that interest you the most (a behavior I also encourage). There are also a few terms that I didn't use or didn't use often but are common on discussion boards and support groups. I've included this basic glossary to help out in both situations.

31313: This is the item or part number of the LEGO EV3 Home Edition. It's a term you will sometimes see on LEGO robotics discussion boards to clarify which set it is that a builder has on hand. The LEGO Education edition part number is 45544. Although both sets contain LEGO EV3 robots, the building parts and sensors are different for each set. See Chapter 1, "What's in the Box?" and Chapter 2, "What's in the LEGO Education Box?" for more information about what is included with each set.

Axle: An axle is a rod used to connect one or more pieces in the LEGO Technic system. The axle has a plus-shaped cross section and comes in different sizes and colors. Axles can be used to transfer force from a large or medium motor to a wheel or gear. For example, in Chapter 4, "Building Your First Bots," axles are used to transfer motion from the motors to the tank treads to make the Track3r go (see Figure A.1).

FIGURE A.1 A variety of axles from the EV3 Home Edition.

Axle Connectors: This describes a part that is used to connect two or more axles together. The part can be straight (turns two axles into one longer, straight axle) or bent (allows force to be transferred at an angle). Angled axle connectors are also known as *angle elements* (see Figure A.2).

FIGURE A.2 The axle connectors in the EV3 Home Edition are all red.

Ball Caster: The ball caster, shown in Figure A.3, is a part found in the LEGO Education set. It consists of a large metal ball bearing and a cup that goes around it. It's used to make a robot that is more maneuverable, much like the castor wheels on shopping carts and office chairs make the chair easier to move. You can find more information and examples in Chapter 5, "Building the LEGO Education Bots."

FIGURE A.3 Here the ball caster is assembled.

Ball Joint/Tow Ball: This is a round ball on the end of some axles, pegs, or other parts. Ball joints are coupled with the sockets in steering gear or track rods to form a very flexible joint, sort of like your shoulder or hip joint. Read more in Chapter 1 (see Figure A.4).

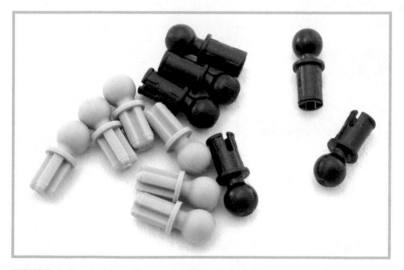

FIGURE A.4 Ball joints with both axle and peg connections.

Blocks/Programming Blocks: Blocks are the basic programming elements in the graphic EV3 programming environment. Blocks could control sensors, variables, or be user defined. Blocks do not refer to the physical LEGO pieces, which are called bricks. For more information on blocks, see Chapter 7, "Make Your First EV3 Program."

Boss and Pin: The boss and pin piece looks a little like a crank with a handle. This piece could be used for that purpose, or it could be used to transfer motion in different ways, such as a piston engine design. See www.technicopedia.com for some fantastic examples of piston and other motors built using LEGO Technic pieces (see Figure A.5).

FIGURE A.5   Boss and pins from LEGO Education.

Bushings: Bushings are end caps for axles. They're mainly used to keep parts such as wheels and gears from sliding off the end of the axle. Learn more in Chapter 1 (see Figure A.6).

FIGURE A.6   Full and half bushings.

Canvas: The programming canvas in the EV3 Home Edition software is the main area for creating programs. Blocks are dragged into sequence on the canvas. For more information see Chapter 7.

Car Parts/Modeling Elements: Car parts or modeling elements are larger pieces designed for both form and function. They're generally placed around the outside of robots, and the EV3 Home Edition provides stickers for decorating these elements. See more in Chapter 1 (see Figure A.7).

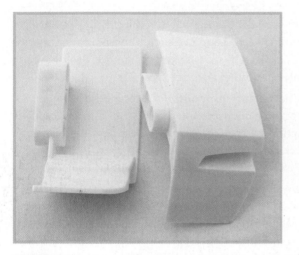

FIGURE A.7 Here are the car parts before stickers are applied.

Caster Wheel: A caster wheel is the wheel equivalent of the ball caster. It is a wheel allowed to spin freely along the horizontal axis. Caster wheels are found on the bottom of shopping carts and office chairs. EV3 sets do not come with caster wheels. They must be created. See Chapter 7 for instructions.

Constant Velocity Joint/CV Joint: The CV joint is shown in Figure 32 of Chapter 2. CV joints consist of two parts that join together to transfer motion at an angle, such as in a robotic arm or engine.

Content Editor: The Content Editor is a digital notebook inside the EV3 Home Edition software that you can use to create notes to yourself or instructions for others. Read more in Chapter 7. The LEGO Education software also includes a content editor, but this is designed for educator use to create or modify lesson activities.

Core Set/Expansion Set: The LEGO Education edition of the EV3 comes in two parts. The core set includes the Intelligent Brick, sensors, and Technic parts for building. The expansion set includes a lot more parts, including more wheels, more tires, and more frame beams. For more information, see Chapter 2.

Cross Blocks: Cross blocks, shown in Figure A.8, are a special type of beam with adjacent perpendicular peg or axle connections. See Chapter 1 for more information.

FIGURE A.8   Two simple cross blocks.

Daisy Chain: Daisy chains are robots formed by connecting one or more Intelligent Bricks together to combine the computing power and sensor ports of each. Programs must specify daisy-chain mode. Up to four Intelligent Bricks can be daisy chained together. See Chapter 11, "Daisy-Chaining Projects" for more information and project ideas.

Data Wire: A data wire connects information (data) from one block to another. For example, the output of a variable block could be connected to the input of a motor block. The sequence wire performs a similar function for the sequence of blocks and allows groups of blocks to be separated from each other on the canvas. See Chapter 8, "More MINDSTORMS Programming: The Line-Following Robot," for more information.

Differential gear: A differential gear (*differential* for short) is a part that is most often used as a motor for wheels that need to turn at different rates of speed. Why would wheels need to turn at two different speeds? When driving around a curve, the wheel on the outside of the curve should actually spin much faster than the wheel on the inside of the curve. In most small robots, this won't make a huge difference, but adding a differential can give your robot more control. See Chapter 2 for more information (see Figure A.9).

FIGURE A.9   A differential gear.

Educator Vehicle: The LEGO Educator Vehicle is a basic design for the LEGO Education version of the EV3. See Chapter 5, "Building the LEGO Education Bots," for more information on the LEGO Education version or Chapter 6, "Hacking What You Have," for a version that can be created from the EV3 Home Edition (see Figure A.10).

FIGURE A.10 The LEGO Educator Vehicle with a gyro sensor and ultrasonic sensor attached.

Enchanting: Enchanting is an alternate, open source MINDSTORMS programming language for the NXT. It was a modification of the Scratch programming language. As this is written, there is no equivalent version for EV3.

EV3: The MINDSTORMS EV3 is a programmable robot by LEGO and part of the LEGO Technic family.

Frames: Frames or beam frames are rectangular beams that allow for large, load-bearing projects. See Chapter 1 for more information (see Figure A.11).

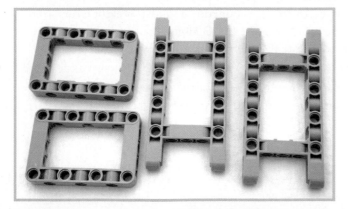

FIGURE A.11 Here are the two basic types of beam frames.

Gear Rack: A gear rack is designed to work with a gear to convert the circular motion of the gear to the linear motion of the rack. See Chapter 2 for more information (see Figure A.12).

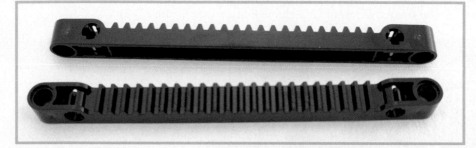

FIGURE A.12 A gear rack can be used to create a lifting or sliding motion.

Gear Ratio: When two or more gears are connected, the difference in the number of teeth reduced to the common denominator is the gear ratio. For example, a gear with 4 teeth and a gear with 12 teeth is a 1:3 ratio. The gear ratio makes a difference, because the gear with the fewest teeth rotates the fastest. That means in a 1:3 ratio, that gear with 4 teeth rotates three times as fast as the gear with 12 teeth.

GitHub: GitHub is a common way open source software is distributed and maintained. Programmers like it because it allows them to keep control over changes and the latest versions of software projects, but you don't have to be a developer or part of the project in order to use GitHub to download software. Alternate programming languages and operating system hacks for the EV3 may be distributed using GitHub. Go to www.github. com for more information.

Home Edition: The EV3 is divided into the Home Edition and the LEGO Education version. Both sets have slightly different parts but are compatible with each other. For more information, see Chapters 1–4.

IDE (Integrated Development Environment): This is a piece of software that makes writing programs easier. Often IDEs allow you to compile your code and help prevent errors from things like typos while you complete your code. The LEGO Home Edition software is an IDE, but you may want to use a separate IDE (such as Eclipse) if you want to use an alternate programming language.

Intelligent Brick: This is the brains and computing heart of the LEGO EV3. The Intelligent Brick has the screen, the sensor ports, the buttons, the processor, the card slot, and the speakers (see Figure A.13).

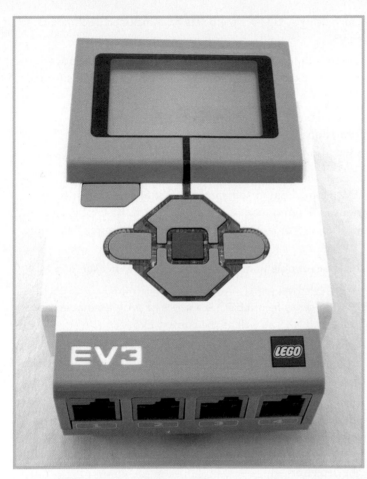

FIGURE A.13 The Intelligent Brick does all the processing for the robot.

Java: Java is a programming language from Sun Microsystems. It's designed to be written once and then run on multiple types of machines using a virtual machine. A port of the Java programming language called leJOS is available for the EV3.

LabVIEW: LabVIEW developed the EV3 computer software.

LEGO Duplo: LEGO Duplo are LEGO bricks designed to be used by toddlers and preschoolers. The parts are larger to prevent swallowing hazards, and they are incompatible with most other LEGO system bricks.

LEGO Education: LEGO Education in North America is a collaboration between LEGO and Pitsco in Pittsburg, Kansas. LEGO Education sells a variety of specialized LEGO systems designed for classroom teachers, including a LEGO Education version of the EV3.

LEGO System: LEGO System bricks are the most common type of LEGO, although most parts are not completely compatible with the EV3. LEGO system bricks typically feature studded blocks that interlock. Increasingly, LEGO system sets have included Technic parts that can also be used with the EV3.

LEGO Technic: The Technic system is designed around motion, motors, and robotics. Rather than using the LEGO system stackable bricks, Technic uses pins and beams for most construction.

leJOS: leJOS is a port of the Java programming language for EV3. See Chapter 12, "Extending Play," for more information.

Linux: Linux is an open-source operating system upon which the EV3 operating system is based.

Lobby: This is the area of the EV3 Home Edition software that first appears when you boot it up on a computer. Various EV3 demo models are shown. From the Lobby, you can either download demo robot instructions or launch a new project. See Chapter 7 for more information.

M: The M measurement is a single beam-hole unit. Beams and axles are measured in M units. See Chapter 1 for more information.

MINDSTORMS: The MINDSTORMS series from LEGO is a series of programmable robots. They include the discontinued RCX and NXT as well as the current EV3.

Mission: EV3 Home Edition instructions for building demo robots are divided into "missions." Each mission has an objective and completes a robot, although some missions lead into each other to complete a larger and fancier robot. See Chapter 4, "Building Your First Bots," for more information.

Mode: On most programming blocks, the bottom-left corner of the block is the mode selector. Modes can control the way the block interacts with the program; for instance, a color sensor port could be used for the modes of color detection, reflected light measurement, and ambient light detection.

NXT 2.0: LEGO MINDSTORMS NXT 2.0 was the predecessor to the EV3 set. See Chapter 3, "Comparing the EV3 and NXT," for more information.

Operating System (OS): This is the program that handles a computer's hardware and software functions and allows it to operate and run other programs. The EV3 uses a Linux-based OS, but alternative operating systems can be loaded onto the EV3, such as leJOS. See Chapter 12 for more information.

Pairing: Pairing establishes a renewable connection between Bluetooth devices. Your EV3 and computer can be paired to transfer EV3 programs to the robot wirelessly.

Pegs/Pins: Pegs (or pins) are round rods that connect Technic beams. Pegs can have more or less friction to allow for easier or slower rotation. See Chapter 1 for more information (see Figure A.14).

FIGURE A.14 A pile of black pegs from the EV3 Home Edition set.

Programming Palette: The programming palette is the lower portion of the EV3 Home Edition software. This area allows you to select blocks and drag them onto the canvas. See Chapter 7 for more information.

RobotC: RobotC is an alternative programming language based on C. The appeal of RobotC is that it includes educational materials and works on a large variety of different robotics systems, including VEX, Arduino, and NXT. A version of RobotC for EV3 was released on August 29, 2014. RobotC is not free, and licenses start at $49 per year.

Sensor: A sensor is a component that can be connected to the EV3 sensor ports using the proprietary flexible cable and performs a particular function for detecting data about the robot's environment. Sensors include the color sensor, the gyro sensor, and the infrared sensor. See Chapters 1 and 2 for more information (see Figure A.15).

FIGURE A.15 The color sensor detects light intensity and color.

Servo: A servo is a motor powered by the EV3. Servos can be connected or disconnected from the EV3 using flexible cable. The EV3 includes two large motor servos and one medium motor servo.

Spikes/Bions/Bionical Eyes: Spikes or bions are small, spikey technic parts that come with the EV3 set. They have one axle connection and no other connection. They can be used in place of bushings in some cases. See Chapter 1 for more information (see Figure A.16).

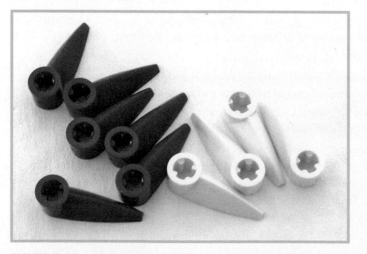

FIGURE A.16 Red and white bions from the EV3 Home Edition.

Stud: A stud is the bump on a LEGO system brick that allows it to connect to other bricks. Studs can also exist as single interlocking bumps and can be used for decorative flourishes on EV3 designs (see Figure A.17).

FIGURE A.17 These are bright red, transparent studs included with the LEGO Education expansion set.

Tension Ring: These are rubber bands that come with EV3 sets and allow you to create projects that require some tension, such as the arms of a robot that should grip tightly. See Chapters 1 and 2 for more information.

Testing Track: The test track comes as part of the EV3 Home Edition box. See Chapter 1 for more information.

Tetrix: Tetrix is a metal building system for larger (relative to the EV3) robots. Tetrix is made by Pitsco, and the system is somewhat compatible with the EV3 system.

Variable: In programming, a variable is a way to handle data that may change over time. Think of it as a bucket that could be filled with a variety of things, such as the name of a person or the last sensor reading. In the EV3 programming language, variables are handled by variable programming blocks.

Wing Parts/Beams with Bows: Wing parts are similar to car parts. In the EV3 Home Edition, wing parts can be decorated with sensors, and in the LEGO Education edition, most wing parts are black. See Chapters 1 and 2 for more information (see Figure A.18).

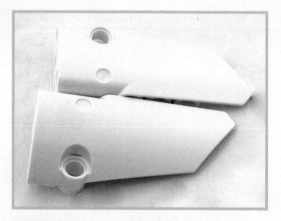

FIGURE A.18 Two wing parts from the EV3 Home Edition.

Worm Gear: The worm gear looks like a screw without a cap. Technically it's a gear that only has one tooth, which makes it easy to calculate gear ratios. The gear ratio is always 1:[*the number of teeth of the other gear*]. So a gear with 20 teeth would have a gear ratio of 20:1, and the worm gear would spin 20 times faster than the large gear (see Figure A.19).

FIGURE A.19 Two worm gears. Notice how there's only one tooth on the gear.

Index

E

U

ultrasonic sensors, 232-234

Undo button, 156

universal joints in expansion set, 63-64

USB cables, 37

USB connections

daisy-chaining with, 77

mini- versus micro-USB connections, 76

on side of bricks, 77

USB slots on EV3 brick, 33

user feedback, programming, 202-206

V

Variable block, 197

variables

calculations with, 200-202

creating, 194-200

velocity joints in expansion set, 64-65

W

WACK3M, 344

Wait block, 195

collision-avoiding robot, 223

wheel assembly

card trick robot, building, 267-268

daisy-chained robot car, 297-307

wheel gears, 20

wheels

caster wheels, building, 141-147

in expansion set, 53-54

moving, 166

in retail EV3 kit, 23

rotation count, changing, 168-169

steering, 167

wheel treads, 23

winding handles, 51

wings

in expansion set, 66

in LEGO Education kit, 44

in retail EV3 kit, 26-27

World Robot Olympiad (WRO), 347

worm gears, 21

writing

documentation, 157

programs, 159-160

changing sensor block modes, 162-164

checking sensor ports, 164

custom blocks, 216-218

custom sounds, 209-210

decision-making with Switch blocks, 172-175

dragging blocks onto canvas, 161-162

flowcharting, 160-161

loops, 175-178, 211-212

moving robots, 165-169

peanut butter and jelly sandwich example, 181-183

switches, 213-216

Timer block, 179

troubleshooting, 206-209

user feedback, 202-206

variables, calculations with, 200-202

variables, creating, 194-200

WRO (World Robot Olympiad), 347

Y

yellow bushings, 16

yellow programming blocks, 158

Z

Znap robot, 119

Zoom button, 156